Sustainable Hard Machining

Sustainable Hard Machining: Implementation and Assessment analyzes the various methodologies of cooling and lubrication employed during hard machining operations, along with their potential contributions towards achieving sustainable machining. It includes the needs, challenges and trends towards sustainable hard machining of difficult-to-cut materials through the application of dry, minimum quantity lubrication (MQL), cryogenic and nanofluid assisted MQL for environmental, economic, ecological and societal benefits, leading to environmentally cleaner sustainable machining.

Features:

- Provides an introduction to hard machining, sustainability and environmentally conscious machining
- Discusses dry and minimum quantity lubrication (MQL) based hard machining
- Includes computational methods and optimization in hard machining
- Reviews nano-cutting fluids in hard machining
- Explores cryogenic cooling in hard machining

This book is aimed at graduate students and researchers in mechanical engineering, manufacturing and materials science.

I0754129

Sustainable Hard Machining

Implementation and Assessment

Ashok Kumar Sahoo, Ramanuj Kumar
and Amlana Panda

CRC Press is an imprint of the
Taylor & Francis Group, an **informa** business

Designed cover image: shutterstock

First edition published 2025
by CRC Press
2385 NW Executive Center Drive, Suite 320, Boca Raton FL 33431

and by CRC Press
4 Park Square, Milton Park, Abingdon, Oxon, OX14 4RN

CRC Press is an imprint of Taylor & Francis Group, LLC

ISBN: 9781032402994 (hbk)
ISBN: 9781032403007 (pbk)
ISBN: 9781003352389 (ebk)

DOI: 10.1201/9781003352389

Typeset in Times
by Newgen Publishing UK

Contents

Preface vii
Acknowledgements ix
About the Authors xi

Chapter 1 Introduction to Hard Machining and Sustainability 1

1.1 Introduction 1
1.2 Sustainable Manufacturing 7
1.3 Sustainable Techniques 12
1.4 Advances in Textured Cutting Tools 41
1.5 Inferences 57
References 60

Chapter 2 Dry and MQL Assisted Hard Machining 71

2.1 Introduction 71
2.2 Machining Performance Under Dry Environment 74
2.3 Machining Performance Under MQL Environment 77
2.4 Sustainability Assessment 80
2.5 Advances in MQL Systems 86
2.6 Inferences 91
References 92

Chapter 3 Cryogenic Assisted Hard Machining 99

3.1 Introduction 99
3.2 Machining Performance Investigations Under Cryogenic Cooling 104
3.3 Sustainability Assessment 120
3.4 Inferences 123
References 124

Chapter 4 Nanofluid Assisted Hard Machining 131

4.1 Introduction 131
4.2 Synthesis, Characterization and Stability of Nano-cutting Fluids 133
4.3 Machining Performance Using Nano-cutting Fluids 147
4.4 Machining Performance Using Hybrid Nano-cutting Fluids 154
4.5 Sustainability Assessment 159
4.6 Inferences 162
References 163

Chapter 5 Computational Methods and Optimization in Sustainable Hard Machining 172

5.1 Introduction 172
5.2 Computational Methods 180
5.3 Multi-response Optimization Techniques 184
5.4 Hybrid Techniques 186
5.5 Inferences 190
References 191

Index 199

Preface

In recent years, hard machining has evolved as an emerging manufacturing process for the benefit of many industries, including reduction of manufacturing cycles, manufacturing costs and setup times, minimization of machine tool usage, elimination of part distortion, minimization of harmful coolant usage due to dry cutting environment, minimization of investment costs and reduction of power consumption as compared to grinding in various applications. It has successively replaced the traditional grinding process and produces comparable surface finish. Sustainable manufacturing has become an important and strategic priority in order to achieve overall production efficiency considering economic, environmental and societal aspects simultaneously. The application of sustainable hard machining has been increasing in various areas of science and technology for use in the aircraft, automotive, defence, aerospace and other advanced industries. Therefore, research is essentially worthwhile to develop its application in industrial sectors. This book provides an overview of hard machining and sustainable manufacturing through machining of various hardened steels and difficult-to-cut materials. *Sustainable Hard Machining* is an indispensable reference/source book for sustainable manufacturing, process design and tool and process optimization, as well as production engineering on shop floors. The book also focuses on various cooling and lubrication strategies for sustainable machining, such as dry, MQL, nanofluid assisted MQL, cryogenic, etc. It will also benefit final-year undergraduate and postgraduate students, as it provides comprehensive information on the hard machining of difficult-to-cut materials to produce high-quality final components. *Sustainable Hard Machining* will also serve as a valuable work of reference for academics, manufacturing and materials researchers, manufacturing and mechanical engineers, and professionals in machining and machine tool technology and related industries.

Acknowledgements

The authors would like to take this opportunity to thank Kalinga Institute of Industrial Technology (KIIT) Deemed-to-be University, Bhubaneswar, India, an Institution of Eminence, for their constant support, motivation and encouragement to accomplish the current work. The authors are thankful to the reviewers, editorial advisory board members, development editor and CRC Press staff for their time and effort on this project. All of their efforts were crucial in producing this book, and we could not have accomplished this milestone without their constant and consistent advice, support and collaboration. Finally, we wish to express our deep sense of gratitude to our parents, family members and friends, who have always stood by us throughout our lives and guided us in times of crisis. This project could not have been completed without our wives' and children's moral support, patience and encouragement. We want to express our gratitude to everyone who took the time to assist us in preparing this book. Above all, we bow to the Almighty, whose blessings and knowledge have guided and assisted us, not only in the course of this project but right throughout our lives.

About the Authors

Dr Ashok Kumar Sahoo, Professor, Mechanical Engineering and Director, R&D (Technology), Kalinga Institute of Industrial Technology (KIIT), Deemed-to-be University, Bhubaneswar, Odisha, India has extensive teaching, research and administrative experience spanning 26 years. He has published more than 200 research articles with h-index 21 (Web of Science), h-index 29 (Scopus), h-index 32 (Google Scholar) and i10-index 87 (Google Scholar) with 2600+ and 3350+ Scopus and Google Scholar Citations respectively. His work has featured in ScienceDirect Top 25 Hottest Articles with seven sponsored research projects from DST, SERB and RPS-AICTE, New Delhi. Dr Sahoo is credited with 15 patents and is an author of the book *Machining of Nanocomposites* published by CRC Press, Taylor & Francis. He has supervised 23 M.Tech. and ten Ph.D. theses. He was listed in the "Top 2% Scientists in the World" and was the recipient of a "Best Staff Award", KIIT University, "Highly Cited Research Award", Elsevier, Outstanding Reviewer award, Elsevier and is a reviewer board member for many reputable journals. He is actively engaged in the research areas of sustainable machining and machinability of advanced materials, hard machining, MQL and nanofluid assisted machining, composite development and machinability, condition monitoring, surface characterization and thin film coatings (CVD and PVD), modelling and optimization, etc. He is a fellow of IE(I) and a life member of IET (UK), ISTE and ISCA.

Dr Ramanuj Kumar is currently working as Associate Professor in the School of Mechanical Engineering, KIIT Deemed-to-be University, Bhubaneswar, Odisha, India. He obtained his bachelor's degree in Mechanical Engineering from KITS Ramtek, Nagpur University, India in 2009 and an M.Tech. in Manufacturing Engineering from NIT Warangal, India in 2012. He obtained his Ph.D. in Mechanical Engineering from the KIIT Deemed-to-be University in 2018. He is a member of the International Association of Advanced Materials and Indian Science Congress. His research areas are hard machining, titanium and nickel based alloy machining, non-conventional machining, MMC, MQL, NFMQL, cryogenics, optimization and soft computing. He has published more than 110 research articles with h-index 19 (Scopus), h-index 23 (Google Scholar) and i10-index 56 (Google Scholar) with 13000+ and 1600+ Scopus and Google Scholar citations respectively. He has published more than 110 international journal and conference papers.

Dr Amlana Panda is an Associate Professor in the School of Mechanical Engineering, KIIT Deemed-to-be University. He received his M.Tech. in Manufacturing Processes and Systems from KIIT Deemed-to-be University in 2009. He obtained his Ph.D. in Mechanical Engineering from the KIIT Deemed-to-be University in 2016. His research interests are focused on sustainable machining processes, tribological analysis in metal cutting, nanofluid MQL and the fabrication and machining of nanocomposites. He has more than 80 research publications in journals and conference proceedings with more than 1200+ citations and an h-index of 19. He is a reviewer for many reputable Journals.

1 Introduction to Hard Machining and Sustainability

1.1 INTRODUCTION

Manufacturing can be simply defined as a value addition process by which raw materials of low utility and value are converted into high utility and valued products with definite dimensions, forms and finish, imparting desired functional ability. Solid state manufacturing processes can be broadly classified into metal forming and metal machining. During metal forming the volume is conserved and shape is achieved through deforming the material plastically in processes like forging, rolling, drawing etc. However, these mostly serve as primary or basic operations for typical products. In around 80% of components produced through metal forming, machining is essentially required to achieve dimensional accuracy, form accuracy and good surface finish in order to achieve the functional requirements.

Hard machining is the machining of materials with hardness of 45–68 HRC using different types of cutting inserts, preferably cubic boron nitride (CBN) and ceramics or similar superhard tool materials. It produces a comparably good surface finish with higher material removal rate as compared to traditional grinding. It is revealed that machining time has been reduced by as much as 60% for conventional hard turning due to the use of lower feed rate and depth of cut [1, 2]. With appropriate selection of geometrical parameters, machining parameters and environmental parameters, a better surface quality can be produced through hard machining than grinding operation. Multiple hard turning operations can be performed in a single setup rather than multiple grinding setups, which yields high accuracy. Typical benefits of hard turning include the reduction of manufacturing cycles and manufacturing costs, reduction of set up time, minimal usage of machine tools, achievable and comparable surface quality, elimination of part distortion, minimization of harmful coolant due to dry environment, minimal investment cost and less power consumption as compared to grinding in various applications [3, 4]. The application of hard machining in industrial applications is shown in Figure 1.1, and qualitative comparison between hard turning and grinding is shown in Figure 1.2 respectively [5].

These days, the combination of hard turning and grinding is performed on a single machine, to great effect, to reduce cycle times. Moreover, the process combination (hard turning and grinding) allows greater flexibility and improved component quality. A typical example of combination machining is the machining of gearwheels

DOI: 10.1201/9781003352389-1

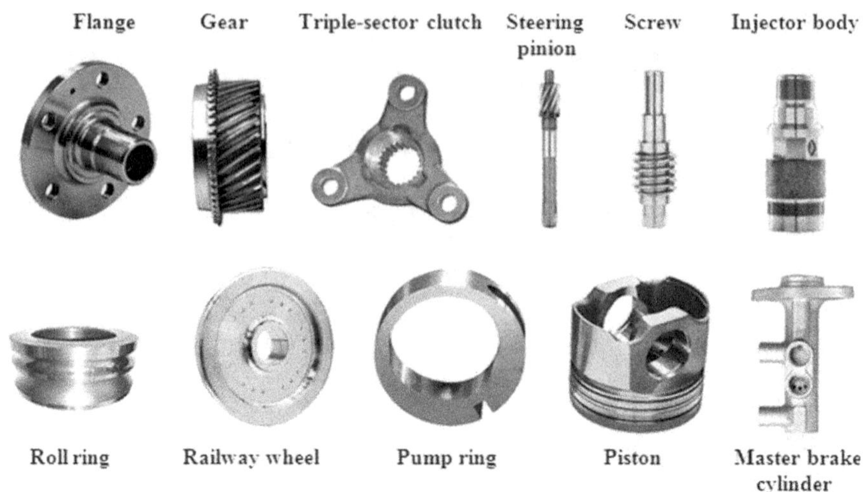

FIGURE 1.1 Industrial applications of hard-part machining. (From Ref. [5].)

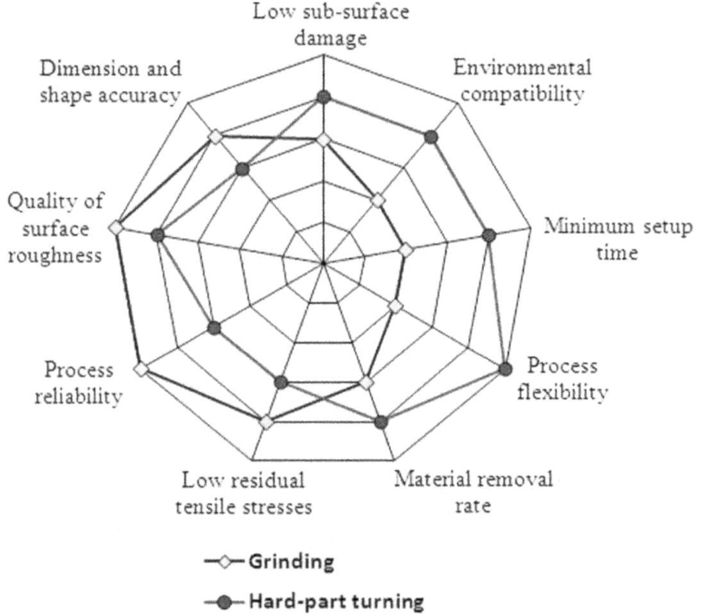

FIGURE 1.2 Qualitative comparison of hard-part turning with grinding. (From Ref. [5].)

as shown in Figure 1.3 (a). The shoulders are finish hard turned and the bore and cone are pre-turned and finish-ground because of the high demands made on their quality [5]. The scroll-free turning process is a process wherein a turning tool carries out a rolling motion which offers new opportunities in the machining of precision components. The cutting edge of an inclined CBN wheel moves over the rotating

(a) (b)

FIGURE 1.3 (a) A combination of hard turning and grinding and (b) scroll-free turning. (From Ref. [5].)

workpiece and the cutting speed is generated by the workpiece rotation, as shown in Figure 1.3 (b) [5].

The desirable properties of cutting tool materials for machining hardened steels are high hardness, high hardness-to-modulus ratio, high abrasive resistance, high thermal conductivity and high thermal and chemical stability. Hard turning is usually performed by a different range of tipped or solid inserts like CBN, polycrystalline cubic boron nitride (PCBN), ceramic tools, cermets and coated carbide [6, 7]. Carbide tools are more popular for machining alloy steels and cast iron. However, these tools possess low hardness but high toughness compared to ceramic and CBN cutting tools. To enhance the surface properties and hardness, thin hard coatings (TiC, TiN, TiCN, Al_2O_3) in single layer, bilayer or multilayer form are deposited over the carbide substrate through physical vapour deposition or chemical vapour deposition techniques. The machining performance and tool life of coated carbide cutting tools are now improved due to the high hardness, wear resistance and chemical stability of coatings compared to conventional cutting tool materials, i.e. high-speed steels, uncoated carbide. Two types of PCBN are used in machining hardened steels, i.e. high CBN content and low CBN content tools. PCBN cutting tools have high hardness, are wear resistant and have high thermal stability but are brittle and thus prone to fracture. So, for implementation of such tools in hard machining, special cutting edge preparation with larger negative rake angle is thus required to improve the strength of the cutting edge.

The following conditions are essential for successful implementation of hard turning [8].

(i) The machine tool should have high rigidity, high surface speed and high accuracy with good surface finish, as the cutting speed during hard turning normally goes as high as 250 m/min.

(ii) Superhard cutting tool materials with high wear resistant properties are essential during hard turning of workpieces of hardness 45 HRC or more because of the generation of higher cutting force. Popular cutting tool materials used by researchers during hard turning are CBN, PCBN, coated CBN tool inserts and ceramics as well as WC coated with TiN or CBN–TiN [9–11].

(iii) Hard turning generates higher cutting temperature as it is operated under dry environment in which reduction of the shear strength of the work material occurs through thermal softening and subsequently favours machining. Therefore, it is advantageous to perform hard machining at higher cutting speed under dry environment. It is economical as well under dry environment.

(iv) The rise of cutting temperature under dry hard machining causes rapid tool wear due to thermal softening and thus deteriorates surface quality and tool life. Raised cutting temperature also causes instant boiling of coolant which reduces the tool life and degrades the surface quality. Application of flood cooling in hard machining also increases the thermal stress/thermal distortions and thus induces the catastrophic failure of cutting tool materials. It is sometimes favoured for the minimization of white layer thickness and so there is always a debate about the application of coolant in hard machining, because of which it is sometimes operated under dry conditions only. But complete absence of coolant drastically affects the tool life and surface quality to some extent [12, 13].

Because of these agreements and disagreements about the application of coolant in hard machining, there is an intermediate methodology of application of coolant called minimum quantity lubrication (MQL) in the cutting zone, or near-dry machining. This will also satisfy the requirement of sustainable hard machining towards achieving environmental, ecological and economic benefits. Hard turning yields white layer formation which affects component service life. Residual tensile stress exists in hard machining that brings the limitation of its application. Hard turning also induces thermal expansion of the cutting tool and workpiece. In spite of these limitations, it has several advantages with respect to time, cost and ecology, which brings attention to the shop floors for finishing of the hardened components such as ball and roller bearings, crank pins and other automotive components and replaces grinding operation. Despite the high potential of hard machining to increase productivity and offset environmental concerns by dry cutting, and its competitiveness with grinding processes, industrial application of this technology is still rather limited. Limitations of hard turning are uncertainties like part accuracy, surface integrity, tool wear pattern, prediction of tool life and economic feasibility. Various limitations of the hard turning process are as follows:

- Tooling cost per unit is notably higher in hard machining compared to grinding.
- Existence of residual tensile stress profiles at the surface.
- Formation of white layer accelerates substantial variations in component performance.
- Errors in dimension, surface roughness and geometric form occurring from tool wear.

Considering these challenges, research in the field of hard turning will definitely be worthwhile. This may lead towards establishing a competitive and economically more effective alternative to grinding.

Appropriate selection of cutting inserts with geometry and cutting parameters during hard machining is a challenge to the machining industries and is considered as a burning issue because of tremendous pressure to achieve good surface quality. Surface quality is greatly affected by the cutting tool wear and heat generation. Thus, there is a need for optimization of cutting parameters, geometrical parameters and environmental parameters to overcome the issues related to hard machining. Appropriate development of prediction modelling is also required in order to increase the machinability of hardened materials. Therefore, research is essential and worthwhile to develop its application in industrial sectors. The requirement of surface quality during hard machining is greatly affected by the amplitude of vibration signals as the radial force component is higher. It strongly affects the surface quality with dimensional tolerances and induces vibrations and chatter. Therefore, the correlation of machining parameters with vibration signals on tool wear and surface roughness during hard machining is worthy of investigation, and these are considered as burning issues nowadays in machining industries for the improvement of productivity.

Sahoo and Sahoo [14] investigated hard turning of AISI 4340 steel using different cutting inserts under dry environment to study the progression of tool wear. The

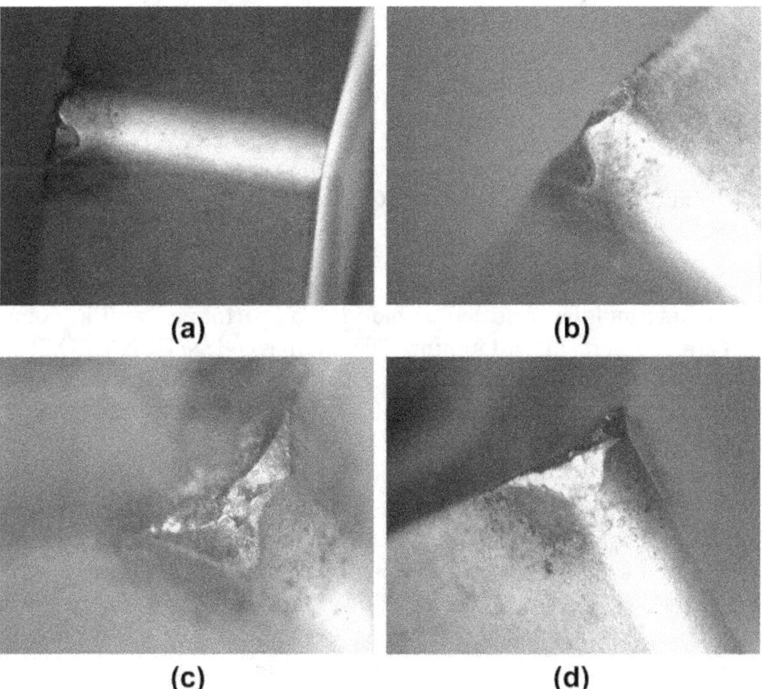

FIGURE 1.4 Tool tips of uncoated carbide insert after machining times (a) 0.46 min, (b) 0.66 min, (c) 0.9 min and (d) 1.77 min. (From Ref. [14].)

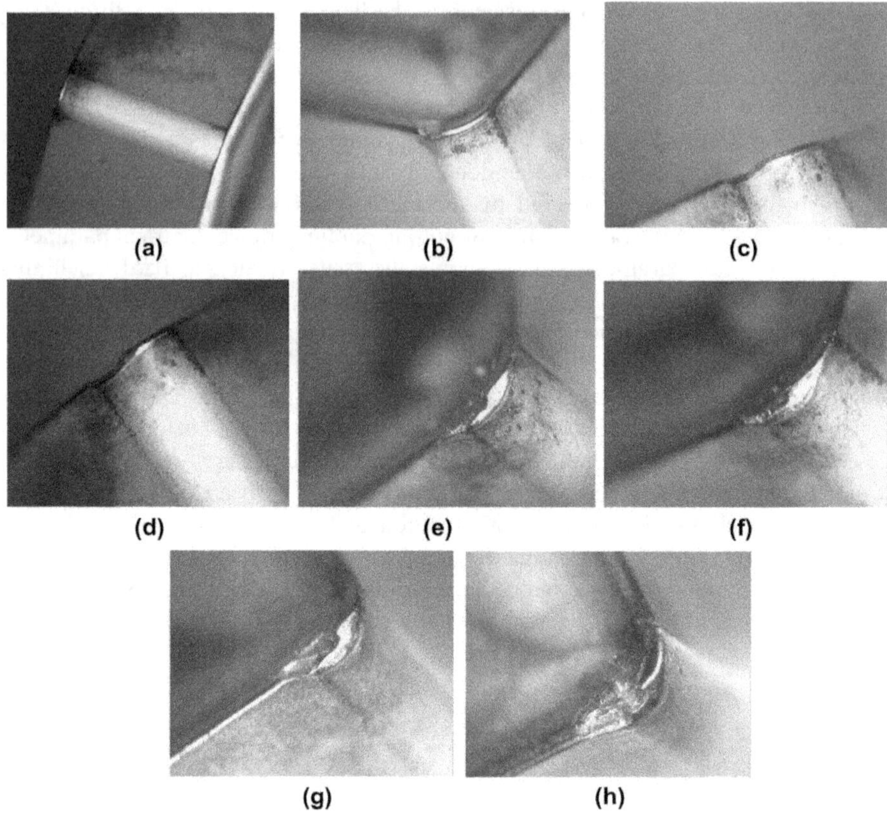

FIGURE 1.5 Tool tips of TiN coated carbide insert after machining times (a) 0.9 min, (b) 1.77 min, (c) 3.44 min, (d) 5.65 min, (e) 8 min, (f) 10.1 min, (g) 15.63 min and (h) 19.48 min. (From Ref. [14].)

results show that multilayer coated carbide inserts performed well as compared to uncoated carbide inserts in hard turning. The multilayer TiN/TiCN/Al_2O_3/TiN coated carbide insert outperformed the uncoated and TiN/TiCN/Al_2O_3/ZrCN coated carbide inserts, having steady growth of flank wear and surface roughness as shown in Figure 1.4, Figure 1.5 and Figure 1.6 respectively [14]. The tool life for TiN and ZrCN coated carbide inserts was found to be approximately 19 min and 8 min, respectively, under the extreme cutting conditions tested. The uncoated carbide insert used to cut hardened steel fractured prematurely. Abrasion, chipping and catastrophic failure are the principal wear mechanisms observed during machining.

Suresh et al. [15] studied the machinability characteristics of hardened AISI 4340 steel using coated carbide insert. Figure 1.7 (a) shows the grooves at the rake surface during machining at a cutting speed of 200 m/min, feed rate of 0.1 mm/rev, machining time of 2 min and 1 mm depth of cut. It may be attributed to the abrasive mechanism and adhesion of the material. Also, it is due to the higher stresses and thermal softening of the tool material at higher cutting temperature in this region. The

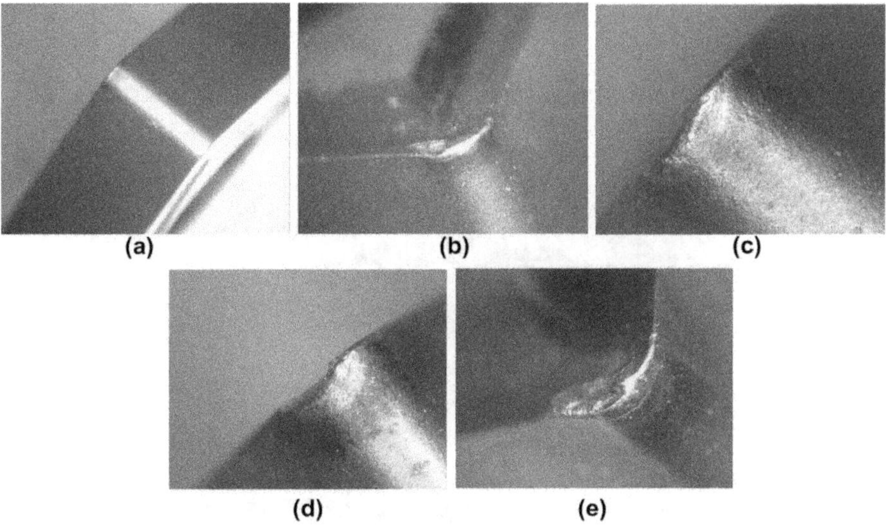

FIGURE 1.6 Tool tips of ZrCN coated carbide insert after machining times (a) 0.9 min, (b) 1.77 min, (c) 3.44 min, (d) 5.65 min and (e) 8 min. (From Ref. [14].)

tool wear mechanism of chipping at the cutting edge is clearly observed in Figure 1.7 (b) at higher parametric conditions such as a cutting speed of 260 m/min, feed rate of 0.26 mm/rev and 1.2 mm depth of cut. Better surface quality is observed at higher cutting speed with lower feed rate. At higher cutting speed with higher depth of cut, short saw-toothed loose arc thick chips were produced, as shown in Figure 1.8 [15].

1.2 SUSTAINABLE MANUFACTURING

1.2.1 Need and Concept

The current advanced manufacturing industries are constantly focusing on alternative machining strategies to yield sustainable goals such as environment-conscious regulations without affecting the material removal rate/productivity [16]. Environmental complexities and economic development go side by side. Manufacturing industries should minimize adverse activities so as to be environmentally benign as per ISO 14000 standards. Therefore, emerging sustainability trends (e.g. the Horizon 2020 target to cut 20% of CO_2 emissions and natural resource consumption) are shifting the industry from non-degradable to biodegradable cutting oils by reducing tons of mineral based fluids (13,726 million tons in 2016 globally with 1% annual increment). Policies are set by the government to meet with requirements for product quality, productivity and environmental pollution. Therefore, there is a need to achieve sustainable manufacturing processes with minimum environmental damage, energy consumption, carbon emissions and machinability characteristics (tool wear, surface roughness and temperature) [17]. In particular, sustainability assessment plays a very crucial perspective before implementation in manufacturing

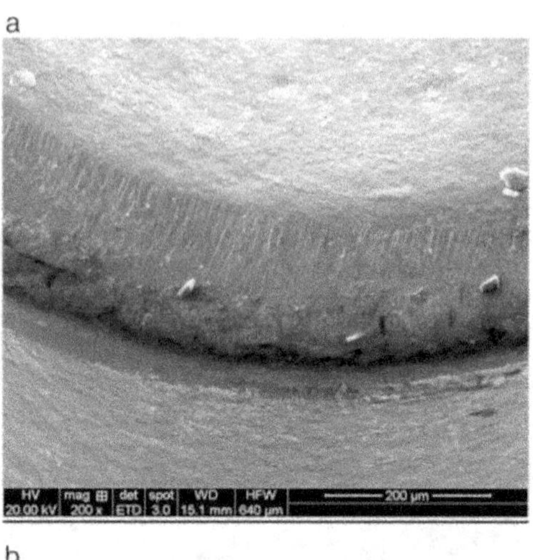

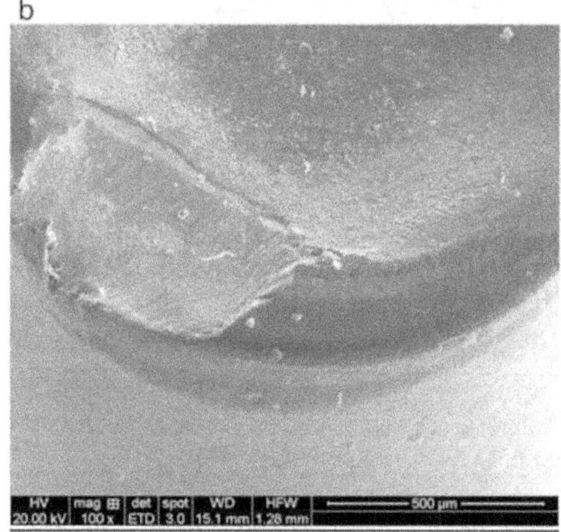

FIGURE 1.7 SEM images of the wear of the cutting tool: (a) v = 200 m/min, f = 0.1 mm/rev, t = 2 min and d = 1 mm; and (b) v = 260 m/min, f = 0.26 mm/rev, t = 2 min and d = 1.2 mm. (From Ref. [15].)

industries for cleaner machining. Sustainable manufacturing implies the manufacturing of quality products with minimum cost, i.e. economical, minimal energy/power consumption etc.

The major users of power and raw materials are the manufacturing industries. Therefore, manufacturing processes should be optimized for power and material requirements so as to have minimal ecological and economic effects. Finished components are produced with optimal usage of resources, using techniques that are economically viable, environmentally benign, ecologically and societally effective and technologically feasible. Energy efficiency with minimal CO_2 emission is the prime requirement of manufacturing industries due to strict government regulations. Therefore, sustainable manufacturing is the current requirement to manufacture

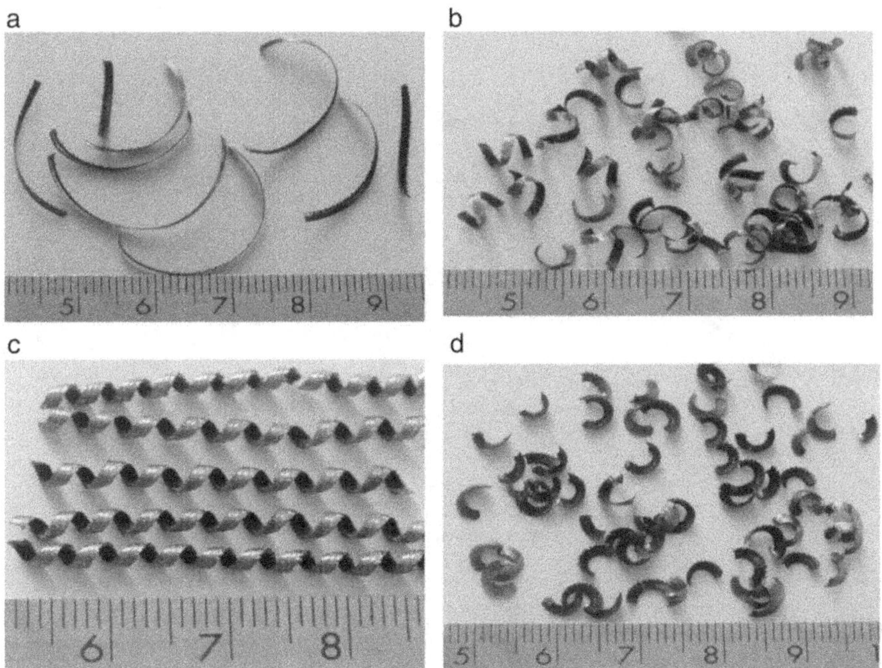

FIGURE 1.8 Aspects of chips obtained as a function of cutting conditions (30× magnification) (a) under cutting conditions v = 140 m/min, f = 0.18 mm/rev and d = 0.8 mm, (b) under cutting conditions v = 260 m/min, f = 0.18 mm/rev and d = 0.8 mm, (c) under cutting conditions v = 200 m/min, f = 0.18 mm/rev and d = 1.2 mm and (d) under cutting conditions v = 260 m/min, f = 0.18 mm/rev and d = 1.2 mm. (From Ref. [15].)

environmentally friendly products with minimal cost, a reduction of waste and maximum utilization of renewable energy sources [18–20]. Sustainability relates to the main three pillars named as i) Economic aspects, ii) Environmental aspects and iii) Social aspects.

There are three steps to achieving sustainable manufacturing: appropriate selection of the work material, sustainable manufacturing process selection and sustainable machining operation. Figure 1.9 shows the pillars of sustainable manufacturing and the steps towards sustainability [18]. Metal machining is considered as the vital industrial sector for producing desired products with accuracy and contributes to the improvement of world economies. Therefore, proper attention should be focused on various aspects such as power consumption, CO_2 emission, tool life, appropriate cutting environments, i.e. cooling/lubrication methods, and higher productivity achievements with profitability so as to be environmentally sustainable with healthy results for the operator. When these are considered, they are mapped with facets of sustainability, the green economy and the circular economy [21]. For technologically effective manufacturing, various factors need to be addressed, such as cost, manufacturing time, surface quality and adaptability. So, the economic,

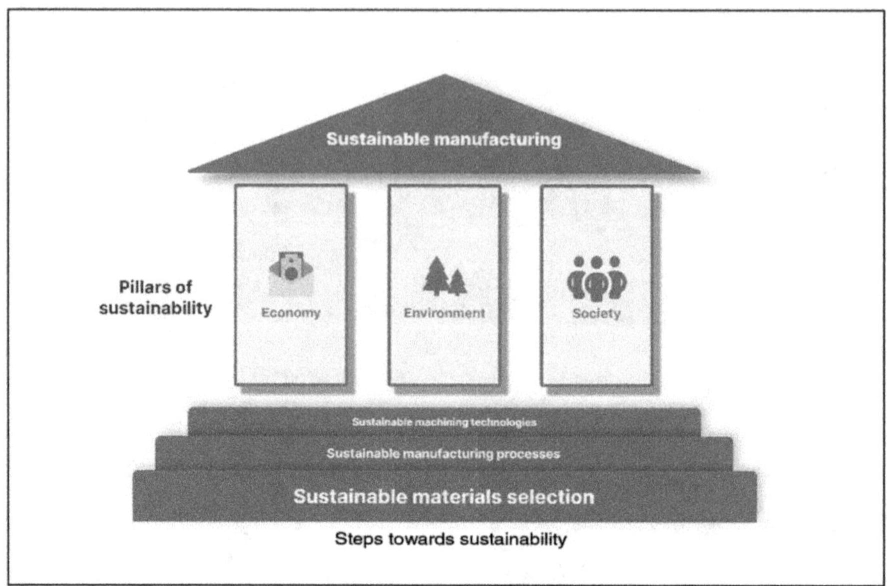

FIGURE 1.9 Pillars of sustainable manufacturing. (From Ref. [18].)

environmental and social aspects of the current manufacturing system need to be considered in order to address environmental consequences and natural resource depletion which is nowadays gaining importance towards sustainability in manufacturing [22]. Sustainable manufacturing refers to an eco-efficient process, i.e. economically efficient and environmentally safe, with social significance, that reduces waste and negative impacts and ensures a hygienic working environment, including worker safety and health, with good product quality, waste management and rates of production, controlling costs and providing training, education and social relationships for workers. Thus, the sustainable manufacturing process can be summarized as reduction of power/energy consumption, waste reduction, improvement of workers' health, improvement of product quality and durability, 3R enhancement (recycling, reuse and re-manufacturing) and development of renewable energy etc.

1.2.2 Sustainable Machining

Three pillars/verticals of sustainable machining relate to the consideration of economic, environmental and societal aspects with the aim of yielding higher profit, a green environment and good social ambience between manufacturer and customer [23]. Maximization of material removal rate and tool life, and less tool wear with minimal machining time are considered economic verticals of sustainable machining. Reduction of power/energy consumption and CO_2 emissions are referred to as the environmental pillar of sustainable machining because machine tool design and the machining process use 40% and 22% of energy consumption, respectively [24].

Satisfaction of the customer implies production of a good quality product with minimal dimensional deviation, and this is termed as the social vertical of sustainable machining. To be precise, it is essential to perform sustainability assessment for cleaner production in the manufacturing industry, not only considering economic aspects but also the effects on society as well as the environment for overall production efficiency. Five major aspects of sustainable machining that should be focused on in the assessment are energy consumption, machining cost, waste management, health and operator safety, and environmental impact. A case study has been shown for assessment of sustainability during machining. Rajan et al. [25] assessed sustainability during the machining of bio-medical titanium alloys under various lubrication environments such as dry, flood and minimum quantity lubrication (MQL) based on a decision-making Pugh matrix approach, energy consumption and CO_2 emission. In the Pugh matrix assessment, some specific weights are allotted to different parameters considering their importance, ranging from –2 to +2 for worst-to-best performance and –1 to +1 for worse and better results, respectively. MQL yields higher scores as compared to flood cooling and dry cutting environment. It is concluded that machining under MQL conditions is observed to be economical and socio-technologically beneficial. Again, MQL machining at optimal parametric conditions reduces energy consumption (by 53.96%) and enhances carbon footprint savings (68.46 kg of CO_2), thus saving on manufacturing costs. Environmentally conscious MQL machining provides better economic and socio-technological benefits towards green machining and cleaner production for industry as per sustainability concerns. Thus, to improve sustainability in machining processes, it is essential to minimize material and energy consumption and pollution as far as the economic and ecological view is concerned [25].

Application of cutting fluids during machining increases cost and environmental impact, and strict protective laws and health regulations put tremendous pressure on the elimination of cutting fluids in machining. In brief, cutting fluid is associated with economic, environmental and health concerns during its application in machining. Dry machining is preferred as the best approach to minimize the effects of cutting fluids related to machining costs and ecological hazards. However, the tool wear rate is excessive and surface quality is low under dry machining and thus cannot be applied in many machining applications. In order to overcome this situation and improve machinability, a new technique has been developed which is gaining popularity nowadays in machining. It is called environmentally conscious machining, minimum quantity lubrication (MQL) or near-dry machining (NDM), to replace traditional flood cooling and thus reduce cutting fluid consumption and health hazards substantially. Slowly, progress has been made towards application of environmentally friendly nanofluids or nanofluid assisted MQL machining, hybrid nano green cutting fluids, cryogenic machining, chilled air cooling and spray impingement cooling to achieve ecological and economic benefits towards sustainable/cleaner/green machining.

The difficult-to-machine materials are broadly classified into three types: hard materials, ductile materials and non-homogeneous materials. The classification and sub-classification of difficult-to-cut materials and environmentally conscious machining techniques are shown in Figure 1.10 and Figure 1.11 respectively [26].

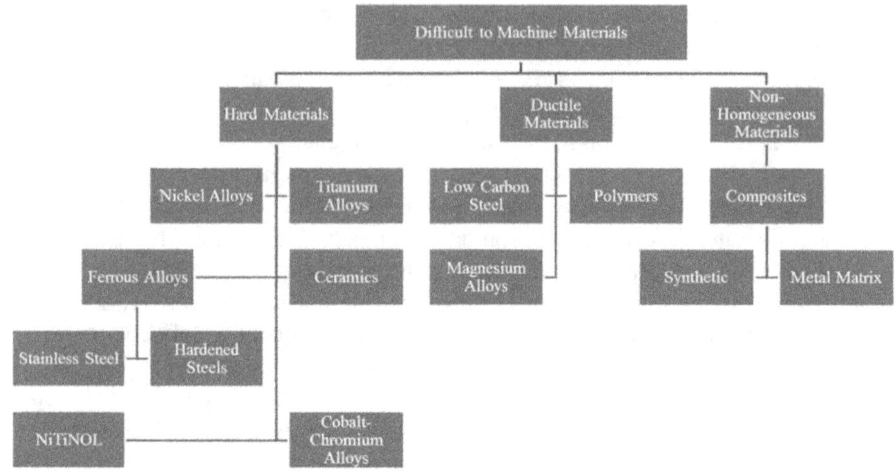

FIGURE 1.10 Classification of the difficult-to-machine materials. (From Ref. [26].)

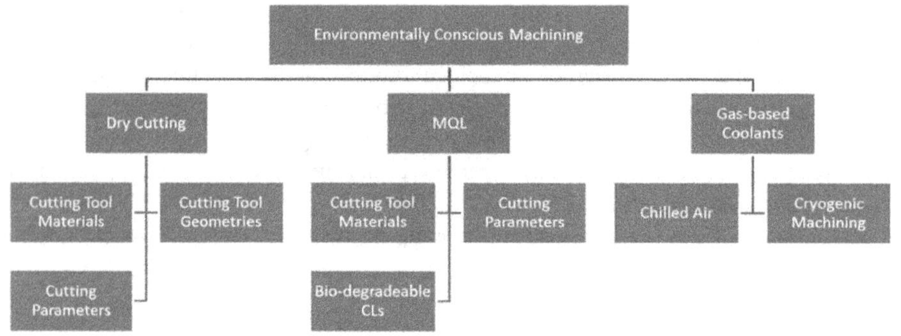

FIGURE 1.11 Classification of different environmentally conscious machining techniques. (From Ref. [26].)

Recently, new cooling/lubrication strategies have evolved, as shown in Figure 1.12, as a ring of sustainable machining [27]. Nanofluid assisted MQL techniques are also evolving as sustainable machining strategies nowadays for the reduction of friction and wear [28].

1.3 SUSTAINABLE TECHNIQUES

1.3.1 Dry Machining

Application of cutting fluids in machining induces health hazards as well as economic and environmental concerns on the shop floor. Dry machining is the preferred choice of manufacturing to eliminate harmful effects of coolants and thus reduce machining costs and ecological hazards. This is considered as the most sustainable means of

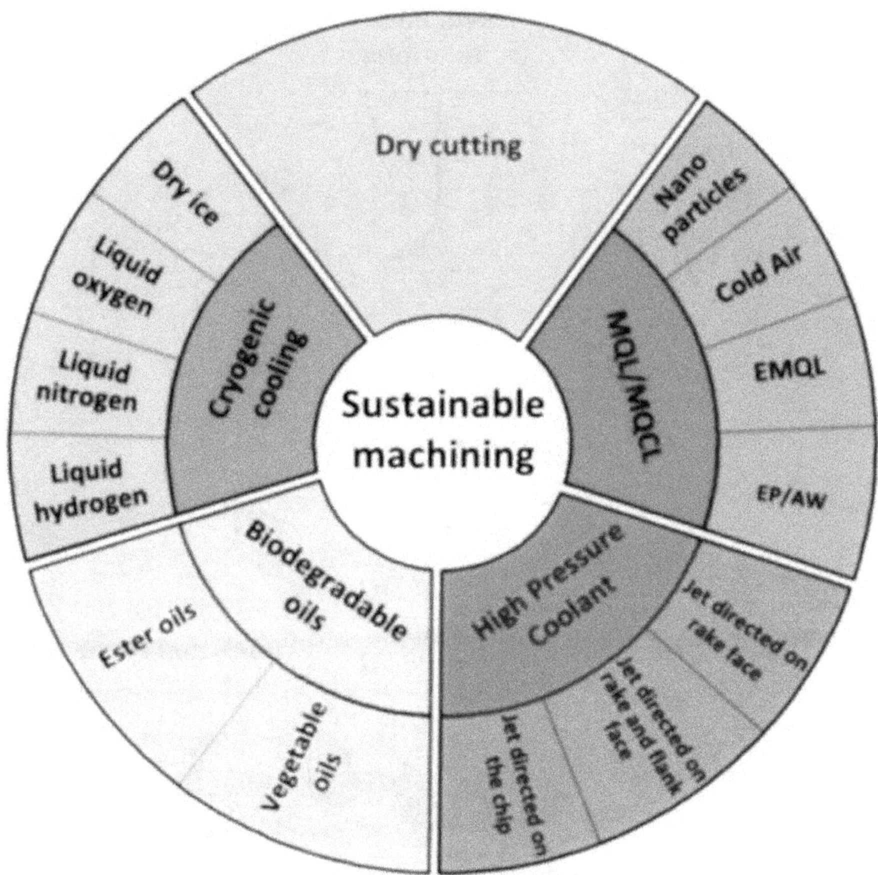

FIGURE 1.12 Ring of sustainable machining. (From Ref. [27].)

machining if machinability is not affected. Dry machining yields higher friction and cutting temperature than wet machining, which affects the tool life, surface quality and dimensional accuracy. Although it may not be the case for all materials, it has some positive effects as well, such as lower thermal shock and cutting cost and enhanced tool life. The advantages of utilization of dry machining are shown in Figure 1.13 [26]. The dry machining performance of cutting inserts can be improved by providing appropriate coatings on carbide substrates. The thin film coating acts as a wear resistant thermal barrier and is lubricious to reduce friction. Again, dry machining performance can be improved by utilizing optimized process parameters in the absence of coolants.

Panda et al. [29] developed correlation models for surface quality characteristics during hard machining of EN31 steel and optimized the machining parameter as well through the Taguchi approach coupled with weighted principal component analysis (WPCA). The most significant parameters affecting surface quality characteristics

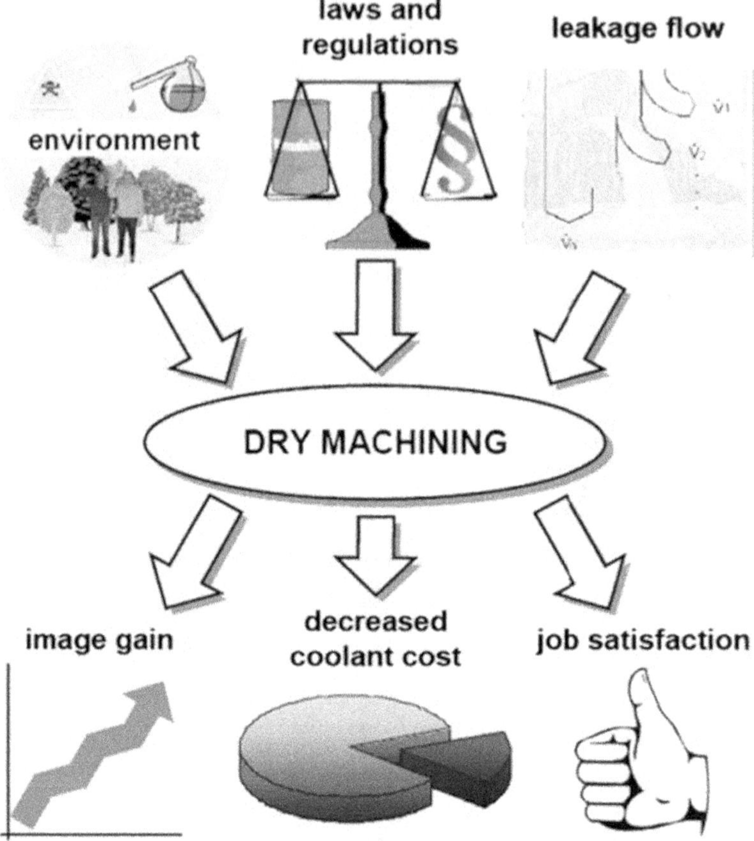

FIGURE 1.13 Benefits of adopting dry machining. (From Ref. [27].)

are feed and depth of cut. Lahres et al. [30] observed that dry machining of 22Mn6 alloy steel replaces the application of flood cooling with the use of TiAlN and Al_2O_3 coatings. The tool life has been improved by up to 80% with the application of TiN+ MoS_2 coating. Panda et al. [31] studied the hard machining of AISI 52100 steel using a coated carbide insert and developed a prediction model for flank wear and surface roughness considering machining parameters and vibration signals online. A multiple quadratic regression model is found to be effective and adequate to predict the responses with accuracy. Krolczyk et al. [27] demonstrated that the tool life is reduced by 65% with the application of mineral oils based cooling-lubricant substances with an intermediate Al_2O_3 ceramic layer as compared to dry machining, as shown in Figure 1.14.

Diniz and Micaroni [32] observed that the increase in tool life during dry machining has been obtained with the increase of nose radius from 0.4 mm to 0.8 mm, which may be considered as a possibility for sustainable machining systems. Fernandez-Abia et al. [33] observed the influence of higher cutting speed on cutting tool and

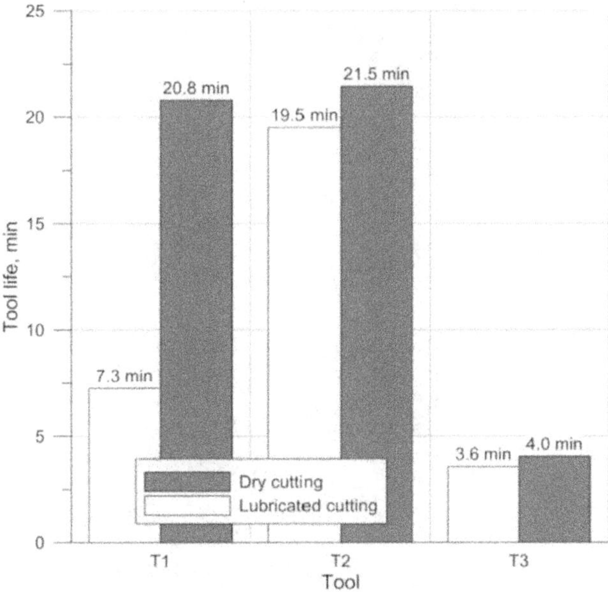

FIGURE 1.14 Effect of cooling on tool life under cut for tools T1, T2 and T3 in the process of duplex stainless steel turning with cooling and without cooling; T1 – rough, T2 and T3 – finish machining. (From Ref. [27].)

workpiece. It is concluded that stainless AISI 303 steel with an austenitic structure is a material recommended for dry cutting. Kumar et al. [34] performed sustainable machining of AISI D2 steel using a coated carbide insert under dry environment and developed response surface methodology (RSM) and artificial neural network (ANN) prediction models for flank wear, surface roughness and cutting temperature. The accuracy of the ANN model is better than that of the RSM model for flank wear, whereas the RSM model is better for surface roughness and cutting temperature. Sahoo and Sahoo [35] conducted hard machining operations using a multilayer coated carbide cutting tool and developed prediction models for responses (flank wear, arithmetic surface roughness average (Ra) and maximum peak-to-valley height within sampling length (Rz)). Model adequacy has been checked and found to be significant as the correlation coefficient is very high. The machining parameters are optimized through grey relational analysis, and considerable reduction of responses are noticed at optimal conditions. An economic feasibility study has been carried out through Gilbert's approach for application of coated carbide inserts in dry hard machining. Total machining cost per part is found to be minimum due to the higher life of the cutting tool, and thus savings are increased by reducing downtime. It yields the economic feasibility of the utilization of carbide cutting tools in machining hardened steel. The tool life has been obtained to be 47 min under optimal parametric conditions during hard turning, and the images of flank wear with successive machining time are shown in Figure 1.15 [35].

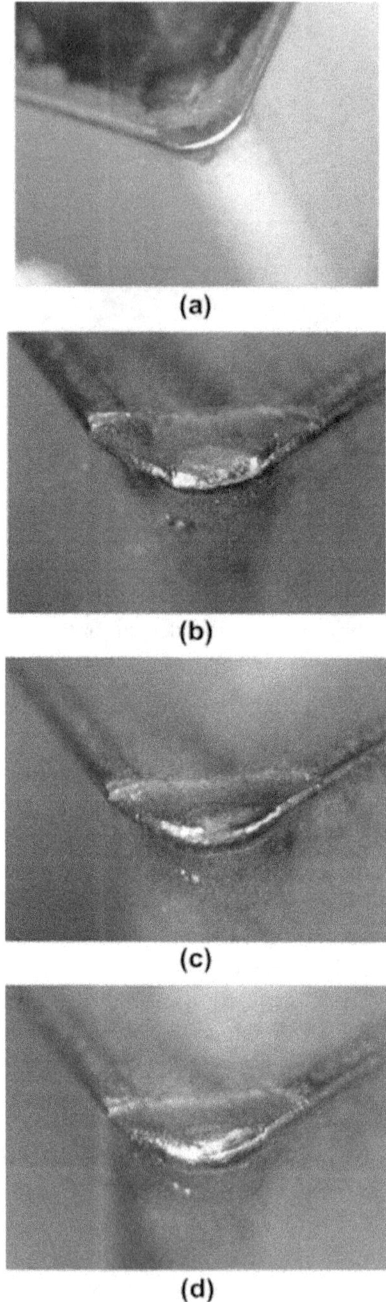

FIGURE 1.15 Images of cutting edge at optimal levels in hard turning after machining times (a) 14 min, (b) 26 min (c) 35 min and (d) 47 min. (From Ref. [35].)

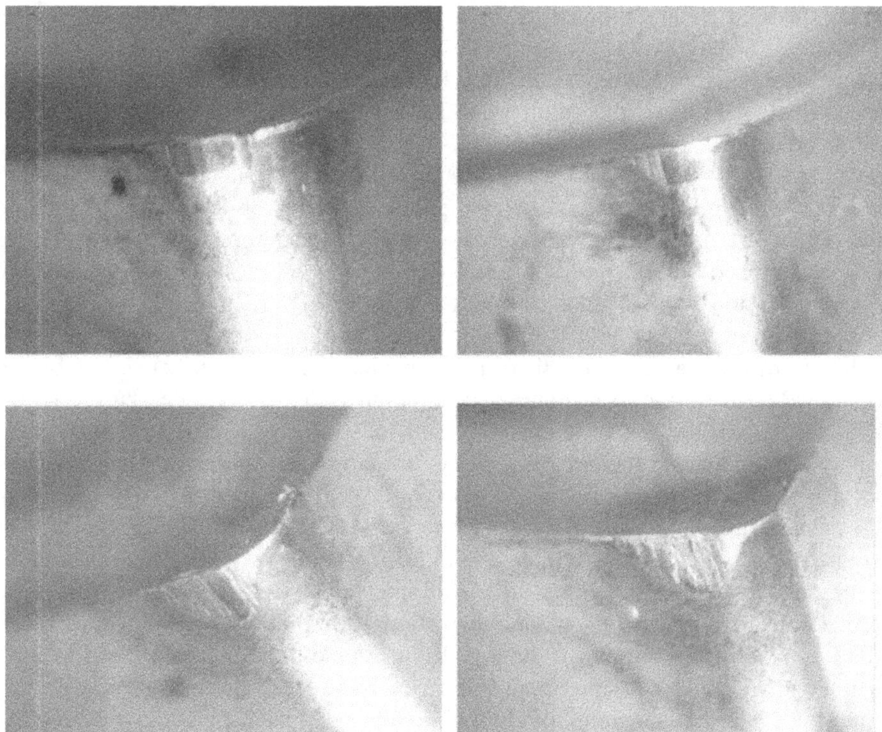

FIGURE 1.16 (a–d) Images of flank wear using uncoated carbide inserts. (From Ref. [36].)

Sahoo and Sahoo [36] conducted a comparative economic feasibility study during dry machining of AISI D2 steel using uncoated and TiN coated carbide cutting tools by estimating total machining cost per part. Cost analysis indicates a 10.5 times higher machining cost per part using an uncoated carbide insert compared to coated carbide during dry machining and thus the coated insert yields 90.5% cost savings. Rapid growth of tool wear has been observed for the uncoated carbide insert, as shown in Figure 1.16, whereas steady growth of tool wear occurred in the TiN coated carbide insert [36]. The tool life for the TiN coated carbide insert has been found to be 30 min longer than that for the uncoated carbide insert under similar parametric conditions.

Park [37] investigated machining of hardened steel using PCBN and ceramic insert without coolant. It was observed that the radial force is the largest component irrespective of the tool used. Cutting force and surface roughness with PCBN tools are both higher and better than ceramic tools in turning hardened steel under similar cutting conditions. Feed rate has the greater effect.

1.3.2 Minimum Quantity Lubrication

Minimum quantity lubrication (MQL) is an alternative technique over flood cooling and is considered as near-dry machining. The minimal amount of mixture of oil and compressed air are sprayed to the cutting zone through an MQL nozzle with a flow rate of 10–100ml/h. The purpose is to lubricate the chip–tool interface so as to reduce the cutting temperature and thus increase the machining efficiency and surface quality. Most of the commercial MQL systems consist of air compressor, coolant containers, tubings, flow control system and spray nozzles etc. The mechanism of all MQL systems is that the pressurized air and coolant are mixed together and a controlled mist is passed through the tube to the nozzle to spray at the cutting zone. The spray nozzle may be external or internal, and sometimes multi-nozzles are also equipped within the MQL system to spray at both flank and rake surface of the cutting tool. This is called a multi-nozzle MQL system. MQL machining is utilized where dry machining is absent or flood cooling is not acceptable. During machining of ductile materials, there is a tendency for adhesion between tool and chip particles due to the rise of cutting temperature, and this is called built-up-edge (BUE) formation. The rise of temperature is mainly due to plastic deformation at the primary shear zone and secondary shear deformation due to friction at the chip–tool interface of the flowing chips. The presence of BUE is detrimental to the machining as it deteriorates the surface quality and reduces the tool life. There is an abrupt increase or decrease of cutting force due to fluctuation of rake angles, which consequently induces chatter and vibration. Variation of workpiece temperature may cause thermal deformation and dimensional deviations. Aluminium alloys are very prone to adhesion and BUE during machining. To overcome the difficulties related to dry machining, MQL plays an important role to cool and lubricate the machining zone. Some studies during machining of aluminium alloys revealed an increase of tool life of up to eight times [38].

Machining of hard materials is usually performed under dry environment due to the ease of shearing and chip deformation at high cutting temperature. Also, as the chips are moving at high speed at the rake surface due to higher cutting speed, the application of coolant becomes redundant. Complete absence of coolant may create difficulties in chip transportation and increases tool wear due to increase of interface friction and cutting temperature. Although application of coolant helps to reduce friction, it increases the machining cost as well. Due to the rise of cutting temperature in hard machining, use of coolant induces thermal stress, and consequently catastrophic failure of the cutting tool takes place. Also, cutting fluid application induces serious health issues during machining, as well as the increase of manufacturing cost, and thus is environmentally not suitable. Therefore, in order to address these issues, a novel sustainable technique called near-dry machining or MQL may be suitably adopted during hard machining for improvement of machinability as well as providing an environmentally friendly, economically feasible and societally acceptable solution. Many researchers have carried out investigations on sustainable hard machining considering environmental aspects and studied the machinability using MQL techniques. The outcomes are discussed.

Chinchanikar and Choudhury [39] carried out an investigation on hard turning of AISI 4340 steel (54–57 HRC) under both dry and MQL environments through coated carbide tools. The enhancement of tool life has been observed under the MQL technique at higher cutting speeds, especially due to better lubrication and cooling effects. The designed MQL setup with mist flow oil/air is shown in Figure 1.17 [39].

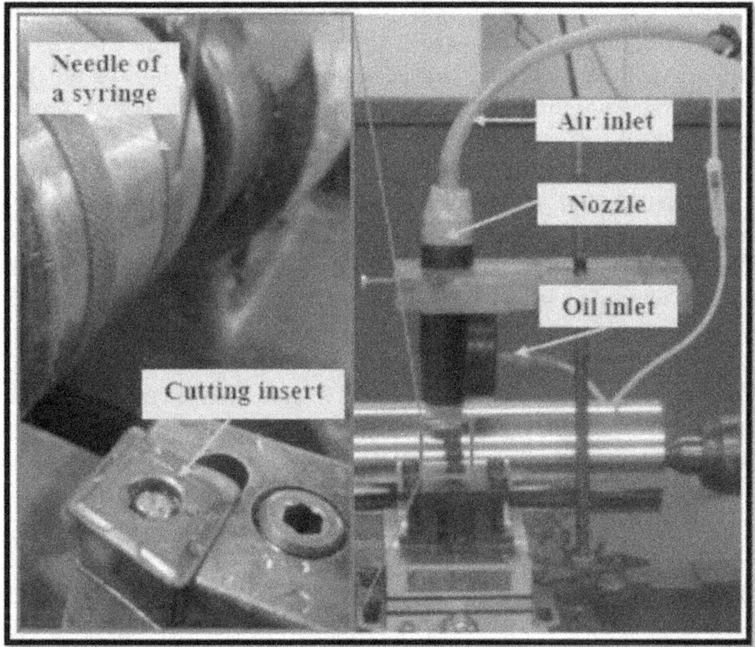

FIGURE 1.17 MQL setup showing view of nozzle having separate inlets for air and cutting fluid. (From Ref. [39].)

Tamang et al. [40] carried out sustainable machining of Inconel 825 under dry and MQL environments. A significant improvement of tool wear, power consumption and surface finish has been obtained under MQL study. Surface roughness values obtained under MQL conditions are found to be low as compared to those in the dry machining environment. For the sustainability of economic, social and environmental aspects, machining parameters are optimized, i.e. minimization of power, surface roughness and flank wear simultaneously through genetic algorithm (GA) and 16 Pareto-optimal fronts solutions. The optimal solution shows better convergent capability for wide application in the machining industries. This solution also provides optimal combination of machining parameters to yield minimal responses. From investigation, MQL is found to be effective with respect to operators' health and environmental consciousness for cleaner machining. Mia et al. [41] investigated hard machining under different cooling-lubrication sustainable environments such as dry, MQL and solid lubricants with compressed air (SL+ CA) and measured surface roughness, tool wear, cutting temperature and chip characteristics. Further, sustainability assessment has been established through a Pugh matrix environmental approach considering parameters such as environmental impact, operator health, cost of coolant, coolant recycling and disposal costs, part cleaning and machining outputs respectively. It is evident that the MQL system improves the machinability during machining and provides environmentally friendly, cleaner machining. Dash et al. [42] investigated surface integrity and chip morphology in machining hardened steel under a nanofluid assisted MQL (NFMQL) cooling environment and also assessed economic and sustainability aspects through a decision-making Pugh matrix tool. Various sustainability assessment parameters were considered for the study, such as worker safety, cutting temperature, surface roughness, environmental impact, coolant cost, recycling and disposal costs of coolant, part cleaning and noise level etc., and scores were assigned to parameters depending upon their importance, ranging from –2 to +2 (inferior to superior results). It is evident from the sustainability analysis and also from the Kiviat radar diagram that machining with NFMQL is found to be socio-technologically effective as well as economically viable. Padhan et al. [43] investigated the machinability and sustainability assessment during machining of austenitic stainless steel under various cooling and lubrication environments such as dry, compressed air (CA), flooded and MQL respectively. A Pugh environmental sustainability approach was utilized to assess sustainability. The MQL environment is considered as environmentally friendly cleaner machining and improved the sustainability during finish machining. Panda et al. [44] studied the machinability of AISI D3 steel using a ceramic insert in terms of cutting force, tool wear and surface roughness by varying process parameters such as approach angle, nose radius, cutting speed, feed and depth of cut respectively under dry environment. Prediction models are developed by multiple regression analysis and adequacy has been checked. Multi-response parametric optimization has been carried out using a hybrid technique of RSM (response surface methodology)–GA (genetic algorithm)–PSO (particle swarm optimization), and this was validated by a confirmation run. Optimal parameters have been used to estimate energy consumption and carbon footprint savings. Finally, sustainability assessment, economic analysis and energy savings were analyzed by Pugh matrix, Gilbert's approach and carbon footprint analysis for cleaner machining/green machining. Abbas [45] investigated machining performance of AISI 1040 steel under MQL nanofluid (vegetable oil mixed with Al_2O_3 nanoparticles), dry and flood environments and obtained optimal parameters which cover both machining outputs and sustainability (CO_2 emission and machining cost) aspects. NFMQL outperformed dry and flood cooling conditions during the machining process, as the best surface quality is obtained due to the improved frictional behaviour of MQL nanofluid mist at the interface. The surface quality utilizing NFMQL has been improved by 34.5% and 85.5% compared to flood cooling and dry cutting, respectively. NFMQL outperformed dry and flood cooling conditions during the machining process as power consumption is reduced and the lowest power consumption has been observed. This may be considered as an effective sustainable method as power consumption and CO_2 emissions are substantially reduced. Dry cutting offers lower total machining cost compared to the other two cooling techniques due to the elimination of costs related to cutting fluids and nanofluids and can be considered as an effective sustainable approach. The lowest machining cost has been obtained in test no. 7 (cutting speed of 150 m/min, depth of cut of 0.25 mm and feed of 0.18 mm/rev) for all cutting environments. Dry cutting reduces machining costs by 7.9% compared to NFMQL application. Abbas et al. [46] investigated sustainability assessment during machining under NFMQL, dry and flood cooling with measured responses of surface roughness and power consumption along with impact on the environment, cost of machining, waste management, safety and health of operators. Sustainability assessment measurements for dry, flood and NFMQL tests along with comparison between all cooling-lubrication based on total weighted sustainability index (TWSI) results are performed. For dry and flood cutting tests, test no. 9 (cutting speed of 150 m/min, feed of 0.06 mm/rev and depth of cut of 0.25 mm) is found to be the optimal sustainable run based on highest TWSI. However, dry cutting offers higher TWSI compared to flood cooling under the same cutting conditions of test no. 9. Test no. 21 (cutting speed of 100 m/min, feed of 0.06 mm/rev and depth of cut of 0.75 mm) is found to be the optimal sustainable run based on highest TWSI for the NFMQL cutting test [46]. Comparison between the most sustainable run for dry, flood and NFMQL tests has been performed. The performance of NFMQL (Al_2O_3 particles suspended in oil–water mixture spray) was observed to be more sustainable, with a TWSI of 0.7 during machining, followed by dry machining with a TWSI of 0.52 and flood coolant with a TWSI of 0.4 [46]. Machinability of MQL nanofluid has been improved due to effective heat transfer and tribological action by the nanoparticle mist that reduces surface roughness and power consumption for sustainability. The optimal parametric settings are found to be cutting speed of 116 m/min, feed of 0.06 mm/rev and depth of cut of 0.25 mm with highest desirability of 0.9050. The minimum surface roughness and power consumption at optimal settings are obtained to be 0.354 μm and 0.528 kW, respectively. Gajrani et al. [47] investigated hard material machining performance through minimum quantity cutting fluids using vegetable based green cutting fluids with additives of solid lubricant nanoparticles. It is observed that 0.3% concentration of solid lubricant MoS_2 based hybrid nano green cutting fluids outperformed other cutting fluids, such as mineral oil, green cutting fluids, 0.3% concentration of CaF_2 based hybrid nano green cutting fluid, with respect to surface roughness (37% better surface finish), cutting force (17% reduction), feed force (28% reduction) and coefficient of friction at chip–tool interface (11% reduction). Sreejith [48] conducted machining of 6061 aluminium alloy and studied the influence of different cutting environments, such as dry, MQL and flooded coolant, on cutting force, surface roughness and tool wear. It is evident that machinability under MQL performed better compared to dry and flood application and can be considered as environmentally friendly, cleaner machining. Masoudi et al. [49], on their machining operation of AISI 1045 steel, concluded that MQL cutting conditions improved the performances compared to dry and wet cooling with respect to surface topography, cylindricity and cutting force. Sustainability assessment performance assessed through a Pugh matrix approach considering environmental impact, operator health and economic and production efficiency revealed the superior performance of MQL machining over dry and wet machining. Goldberg [50] recommended that helical cutting edge geometry and variable pitch flute pattern reduces vibration during machining. The productivity has been improved a lot utilizing the proposed tooling system along with dry or MQL conditions with ecological benefit and minimization of the energy consumption. Upadhyay et al. [51] identified the problems associated with flood cooling application in machining and the need for MQL followed by its working principle. Based on a review of turning and milling, the influence of operating parameter on MQL performance and the effect of MQL on machinability aspects have been presented. Faga et al. [52] investigated the influence of dry and near-dry (such as MQL) environments during machining of titanium alloy on sustainability implications considering tool wear, cutting forces, surface quality, lubricant consumption and health hazards etc. The study also proposes environmental impact minimization along with productivity enhancement. Shokoohi et al. [53] developed an eco-friendly water mixed vegetable oil cutting fluid integrated with antibacterial agent and applied through MQL with a new cooling approach of pre-cooling the workpiece during the machining of hardened AISI 1045 steel. Machining performance has been assessed through responses such as surface roughness, power consumption and chip formation including machining hazards. Significant improvement has been observed in health and ecological concerns as well as performance in machining and control of bacterial growth with the application of vegetable oil as compared to straight oil. Debnath et al. [54] reviewed the development of vegetable oil bio-based cutting fluids and their application in machining and observed the minimization of ecological issues. A brief technique of a dry cutting, MQL and cryogenic cooling cleaner approach in machining has also been presented. A significant reduction of cutting fluids with better machining performance is also noticed compared to the conventional wet cooling approach. Deiab et al. [55] observed improvement in machinability performance through MQL and minimum quantity cooled lubrication environments with rapeseed vegetable oil cutting fluid for sustainability as compared to dry and cryogenic cooling in terms of tool wear, surface roughness and energy consumption, as shown in Figure 1.18 [55], Figure 1.19 [55] and Figure 1.20 [55].

Priarone et al. [56] studied the application of emulsion mist cutting fluids during machining of Ti–48Al–2Cr–2Nb alloy, and improved tool life has been noticed as compared to dry and MQL cutting. Gupta et al. [57] investigated sustainable machining of Al 7075-T6 alloy under different cooling-lubrication techniques such as dry, nitrogen cooling, nitrogen MQL and Ranque–Hilsch vortex tube nitrogen MQL (R-N_2 MQL), respectively. It is evident that R-N_2 MQL performed better compared to other cooling methods, as reduction of surface roughness and tool wear were found to be 77% and 118% respectively and thus enhanced sustainability by saving resources. Pereira et al. [58] investigated machining of Inconel 718 using natural biodegradable oil assisted MQL and compared this with other oils such as sunflower oil, oleic sunflower oil, castor oil and ECO-350 recycled oil. ECO-350 recycled oil is observed to be feasible and increases the life of the cutting tool by 30% compared to canola oil, but the recycling process has to be improved. Considering all issues, oleic sunflower oil improves the tool life by 15% with environmental impact similar to canola oil and thus may be considered as efficient, environmentally friendly and technically viable. Das et al. [59] conducted a machinability investigation of hardened AISI 4340 steel under dry and MQL environments and measured surface roughness and tool wear. Satisfactory performance has been noticed for MQL machining compared to

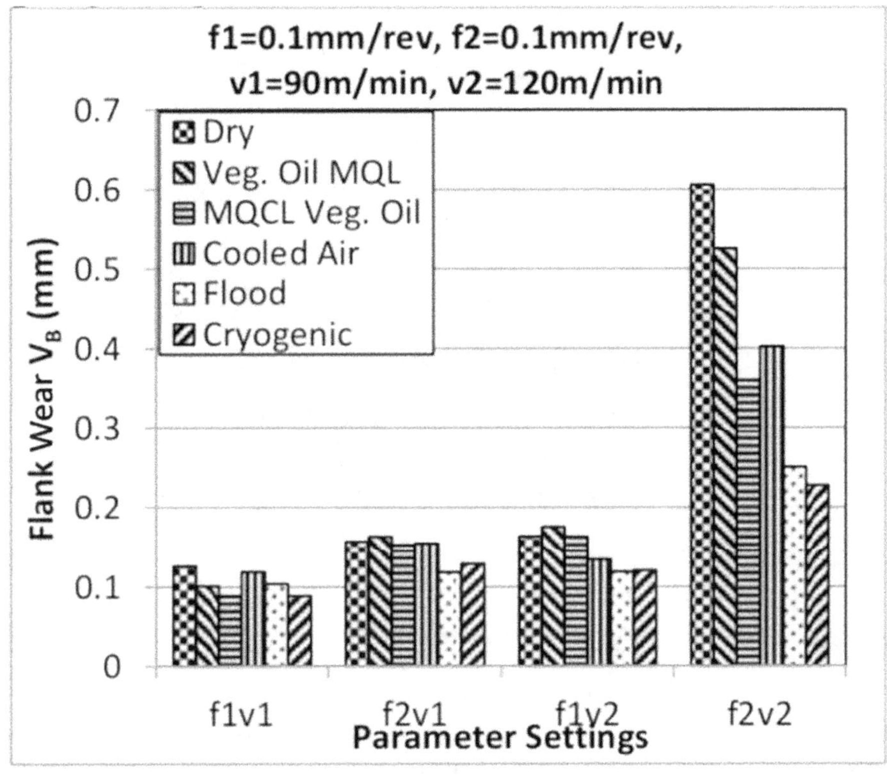

FIGURE 1.18 Tool wear with lubrication techniques. (From Ref. [55].)

dry machining, as good quality product is obtained under MQL environment. Kumar et al. [60], on their hard machining of AISI D2 steel, considered nanofluids (Al_2O_3 and TiO_2 water based) at different concentrations of weight and applied to the cutting zone through air assisted spray impingement cooling using an inexpensive coated carbide insert. Very few investigations into hard machining under nanofluid assisted air-water spray cooling have been performed. Machinability performance is observed to be better using TiO_2 nanofluid compared to Al_2O_3 nanofluid due to better lubricious, wettability and heat dissipation characteristics, particularly at 0.01% weight concentration of TiO_2. At this concentration, there is a 29%, 9.7% and 14.3% reduction of tool wear, cutting temperature and surface roughness noticed, respectively, as compared to the Al_2O_3 nanofluid. Also, a reduction of 29% and 27.7% of tool wear using 0.01% weight concentration of TiO_2 was observed as compared to dry and air-water spray cooling, respectively. Tool life at 0.01% concentration of TiO_2 nanofluid is 2.52 and 1.47 times higher than dry cutting and air-water spray cooling, respectively. Roy et al. [61] reviewed different cooling/lubrication strategies for MQL machining as well as various cutting fluids such as vegetable oil, mineral oil, synthetic oil and nanofluid assisted environments, particularly in machining application for sustainability and environmental, economic and technological benefits.

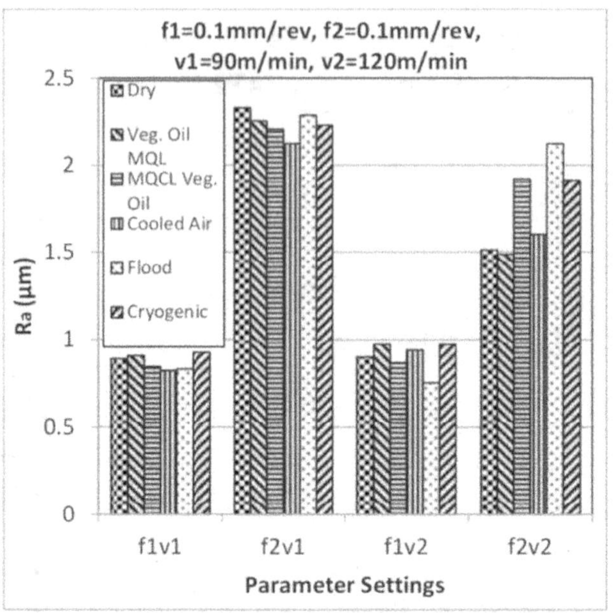

FIGURE 1.19 Surface roughness with lubrication techniques. (From Ref. [55].)

Das et al. [62] investigated comparative machinability performance (tool wear and surface roughness) of hardened AISI 4340 steel under dry and MQL environments. MQL machining is observed to be better compared to dry conditions even at higher cutting speed. Dambhare et al. [63] investigated the sustainability aspects in turning with a case study in the Indian machining industry. The effect of process parameters, cutting environments and types of cutting tools on sustainability factors such as surface roughness, material removal rate and energy consumption was studied. From the analysis of variance (ANOVA), it is revealed that the cutting environment and type of tool affects the surface roughness. Material removal rate is influenced majorly by cutting parameters and type of tool, whereas energy consumption is influenced by cutting environment, type of tool, cutting velocity and depth of cut. Prediction models through RSM and experimental results are very close to each other. The parameters are optimized for sustainability with due weightage/importance to minimum power consumption through a desirability approach. Das et al. [64] investigated machining of hardened AISI 4340 steel under various cooling applications such as compressed air, water-soluble coolant assisted MQL and a nanofluid (Al_2O_3 nanoparticle with eco-friendly radiator coolant) assisted MQL. The nanofluid assisted MQL application showed better machining performance (reduction of flank wear, cutting force and improved surface finish) compared to the other two. The MQL technique has been observed to be effective in minimizing both health risks and machining costs. Haq et al. [65] investigated face milling operation of IN718 (Inconel 718) under two environmentally conscious water based lubricating conditions, MQL and NFMQL (nanofluid assisted MQL), and measured surface roughness, tool-work temperature,

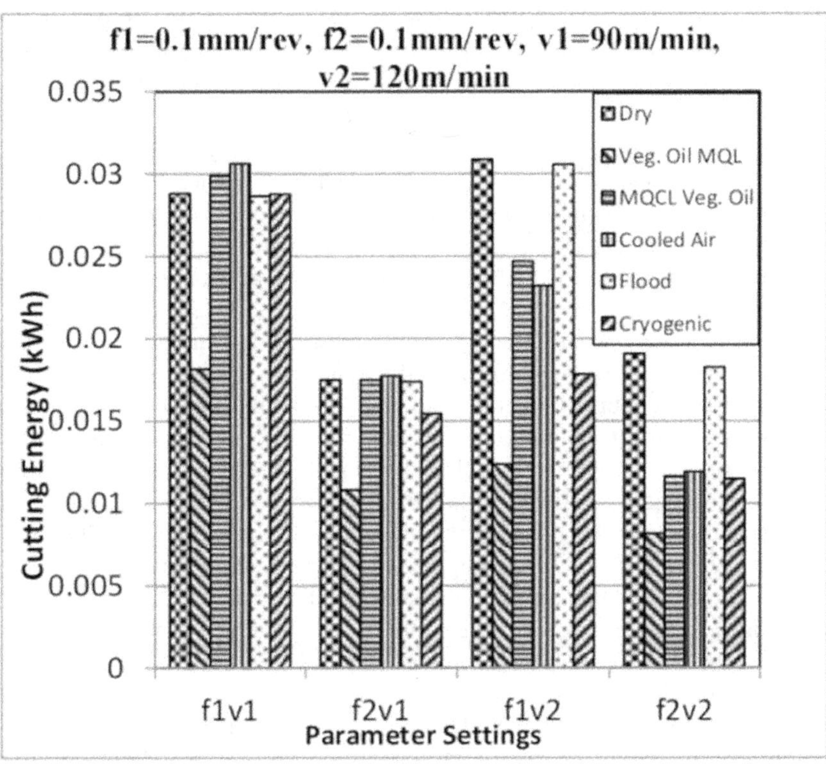

FIGURE 1.20 Cutting energy consumption with lubrication techniques. (From Ref. [55].)

power consumption and material removal rate respectively. A superior sustainability rating has been observed under NFMQL cutting conditions with achievable surface roughness of 0.1 μm or even lower. Davim et al. [66] performed machining of metal matrix composites (MMCs) under an MQL environment at different flow rates and measured surface roughness and cutting power. It is evident from the investigation that the cutting power and surface roughness variation at different flow rates is marginal and is found to be economical. Under MQL conditions, the performance of machining of MMCs has been improved. Tayal et al. [67] investigated sustainable machining of Monel 400 superalloy under dry environment and measured surface roughness, power and cutting force respectively. Feed and cutting speed are the principal factors that affect the surface roughness. The reliability and economics are analyzed for the feasibility of the cutting tool. Increasing cutting speed increases the surface finish and reduces the cutting force in dry machining and thus reduces power consumption. Gupta et al. [68] conducted sustainable machining of Inconel 800 under dry, vegetable oil with MQL, graphene nanofluid with vegetable oil MQL (NFMQL) and liquid nitrogen (LN_2) environments. The performance of LN_2 has been found to be better than other cooling conditions. The reduction of total machining cost per part has been obtained to be 9.3%, energy consumption has been found to be reduced by

11.3%, carbon emission by 49.17% and tool wear by 46.6%. There is an improvement of efficiency of LN_2 by 9.3% as compared to dry machining. Improvement of machining efficiency for MQL and NFMQL have been found to be 2.3% and 6% respectively, as seen from Figure 1.21 [68]. Also, machining cost per part for liquid nitrogen has been reduced by 36.8%, 26.9% and 13.2% as compared to dry, MQL and NFMQL methodologies, respectively, as is evident from Figure 1.22 [68]. From the overall comparison, as seen in Figure 1.23 [68], it is concluded that the improvement of machining performance has been obtained under an LN_2 cooling environment, followed by NFMQL, MQL and dry conditions respectively and is denoted as a sustainable condition.

Das et al. [69] carried out a comparative machinability investigation under dry and MQL assisted hard machining of AISI 4340 steel. MQL assisted machining provides appropriate lubrication at the cutting zone and thus induced better performance compared to dry machining in terms of surface quality, flank wear and dimensional deviation. The progression of flank wear with machining time determines the end of tool life and it is determined when flank wear exceeds the criterion of 0.2 mm. From the experimental observations, the life of the cutting tool under MQL assisted machining increases by 122.15% compared to dry machining. The growth of surface roughness and auxiliary flank wear with machining time has also been undertaken, as auxiliary flank wear affects the surface quality of the product. From the investigation, it is evident that MQL assisted machining induces better surface quality compared to dry machining. The auxiliary flank wear is also significantly reduced by 13.16% under MQL environment as compared to dry cutting. The dimensional deviation takes place during machining due to rapid acceleration of tool wear. The MQL assisted environment provides better lubrication penetration at the cutting zone and reduces the cutting temperature substantially, thus retarding the growth of tool wear. Due to this, dimensional deviation is comparatively reduced as compared to that under dry machining. Also, MQL reduces total machining cost per part and yields 16.21% cost savings over dry machining so is observed to be cost-effective. From investigation, it is concluded that MQL may be considered as a novel alternative over dry and conventional flood cooling because of environmental, ecological and economic benefits and thus may be adopted for cleaner/green machining and industries.

1.3.3 Cryogenic Machining

Cryogenic machining refers to the machining of materials at very low temperature, below 120 K, through a liquefied gases/cryogen medium and directed to the machining zone. The purpose is to minimize the cutting temperature by absorbing the heat from the cutting zone, cooling the tool and workpiece. Liquid nitrogen and liquid helium are considered as common cryogenic coolants used in machining for environmental reasons as they are not pollutants. However, liquid CO_2 can be treated as a pollutant as it has been extracted from the exhaust gases of power plants to minimize contamination. Liquid CO_2 is heavier than air and may cause accumulation and deficiency of oxygen on the shop floor. Due to machining under cryogenic lower temperature, the properties of work and cutting tool materials may change. The cutting

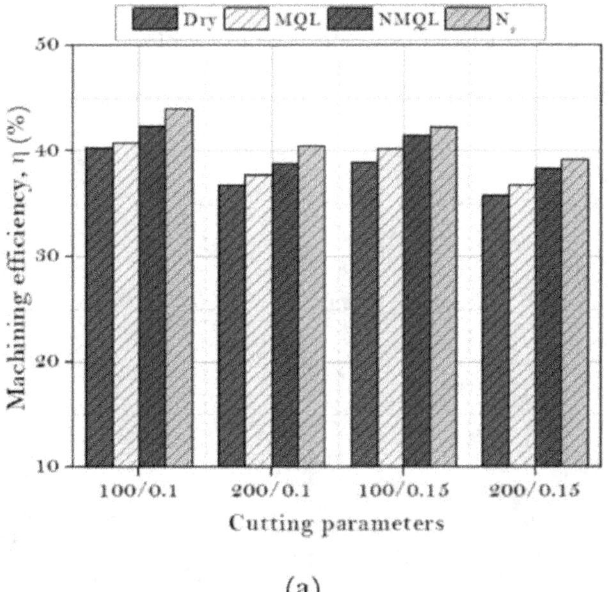

(a)

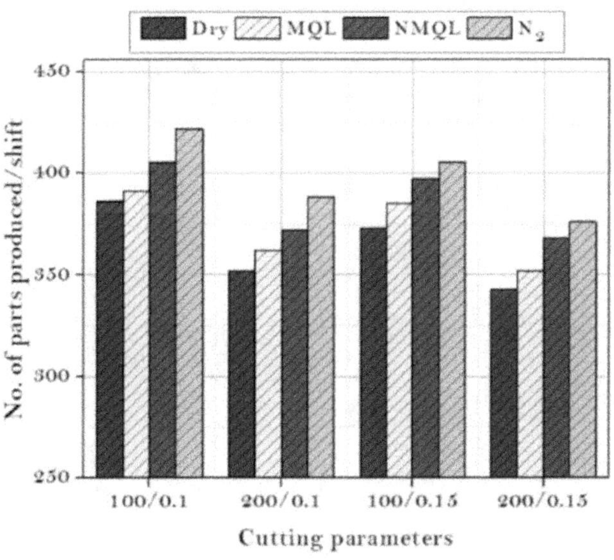

(b)

FIGURE 1.21 Impact of turning parameters (cutting speed/feed rate) and cooling/lubrication conditions on (a) machining efficiency and (b) productivity. (From Ref. [68].)

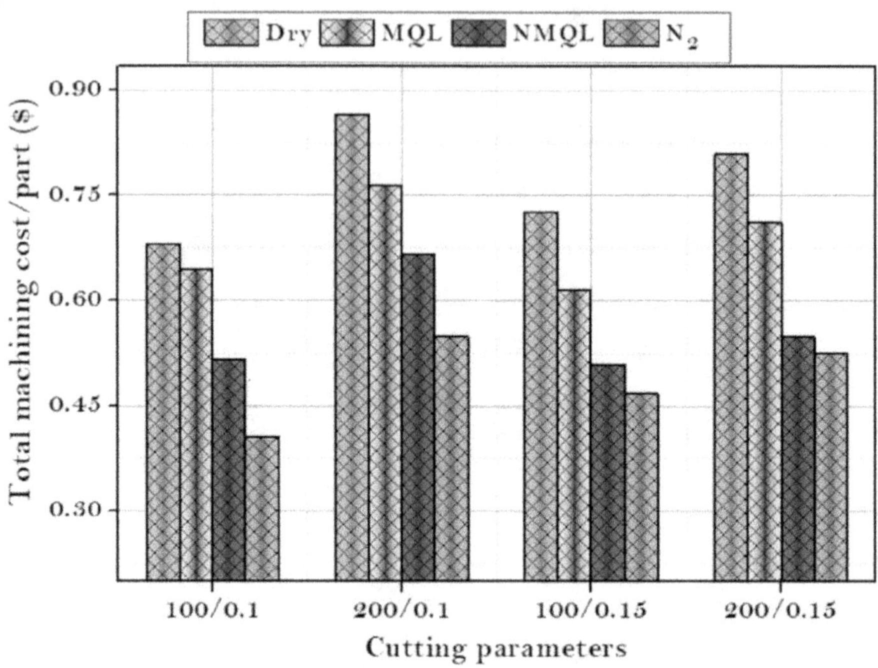

FIGURE 1.22 Impact of turning parameters (cutting speed/feed rate) and cooling/lubrication conditions on total machining cost per part. (From Ref. [68].)

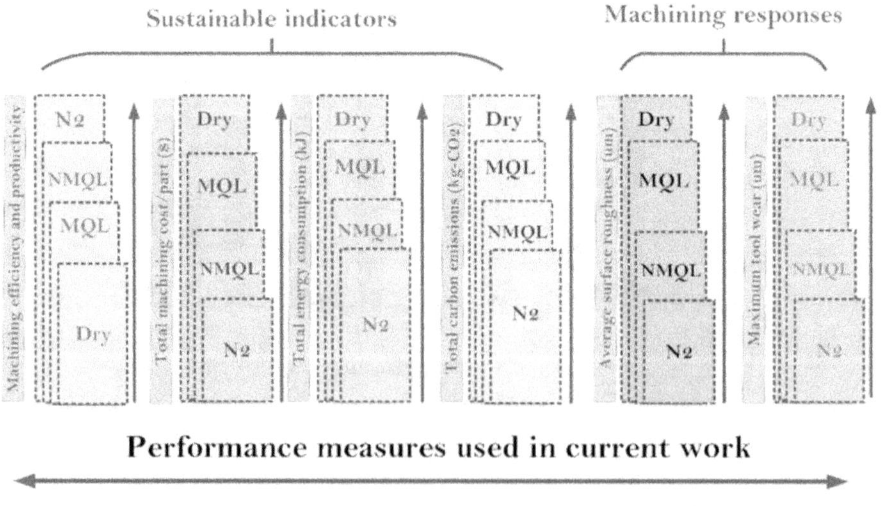

FIGURE 1.23 Overall comparison between all responses. (From Ref. [68].)

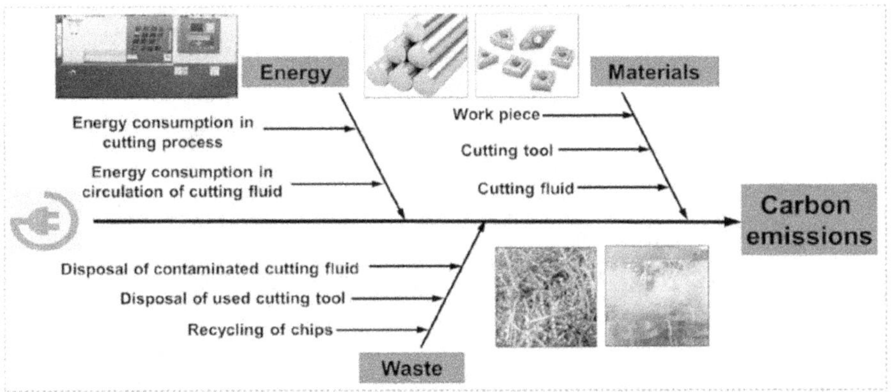

FIGURE 1.24 Causes of carbon emissions in the machining process. (From Ref. [79].)

tool material may undergo increase in hardness and strength and lower the fracture toughness, which in turn enhances the wear resistance of cutting inserts and extends tool life. The most common coolants for cryogenic machining include liquid nitrogen (LN_2), liquid carbon dioxide (LCO_2), solid carbon dioxide (dry ice), liquid helium and air. Reduction of cutting temperature and chip–tool interface reaction occurred by application of cryogenic coolant at the machining zone. This consequently reduces the mechanisms of tool wear such as adhesion, BUE formation, abrasion and diffusion and enhances the surface finish of the components. Venugopal et al. [70] investigated the machining of titanium alloy using LN_2 as a coolant, which significantly reduces the crater and flank wear by 77% and 66%, respectively, as compared to dry machining. There is significant increase of tool life due to the application of LN_2 cryogenic coolant and thus it can be operated at higher cutting speeds [71–74]. It has been observed that LN_2 cryogenic coolant has good lubricating properties in addition to the cooling properties, which decreases the friction at the chip–tool interface/secondary deformation zone. The machining system becomes favourable and is enhanced due to the increase of surface hardness by application of LN_2 coolant that brings changes of coefficient of friction and force of friction [75–78].

Agrawal et al. [79] conducted machining of titanium alloy under cryogenic and wet cutting environments for sustainability. Causes of carbon emissions in the machining process have been illustrated in Figure 1.24 [79]. The carbon emission reduction has been found to be 22% for a cryogenic environment compared to wet machining at higher cutting speed, as evident from Figure 1.25 [79], and total machining cost is reduced by 27%, as shown in Figure 1.26 [79]. Cryogenic cooling has been noticed to be more sustainable during machining than wet machining of titanium alloy.

Salvi et al. [18] investigated the machinability of wrought and additive manufactured (AMed) Inconel 625 (IN625) under sustainable machining environments such as dry, cryogenic (LCO_2) and electrostatic minimum quantity lubrication (EMQL). The total carbon emissions (CE) have increased by 123.95%, 79.43% and

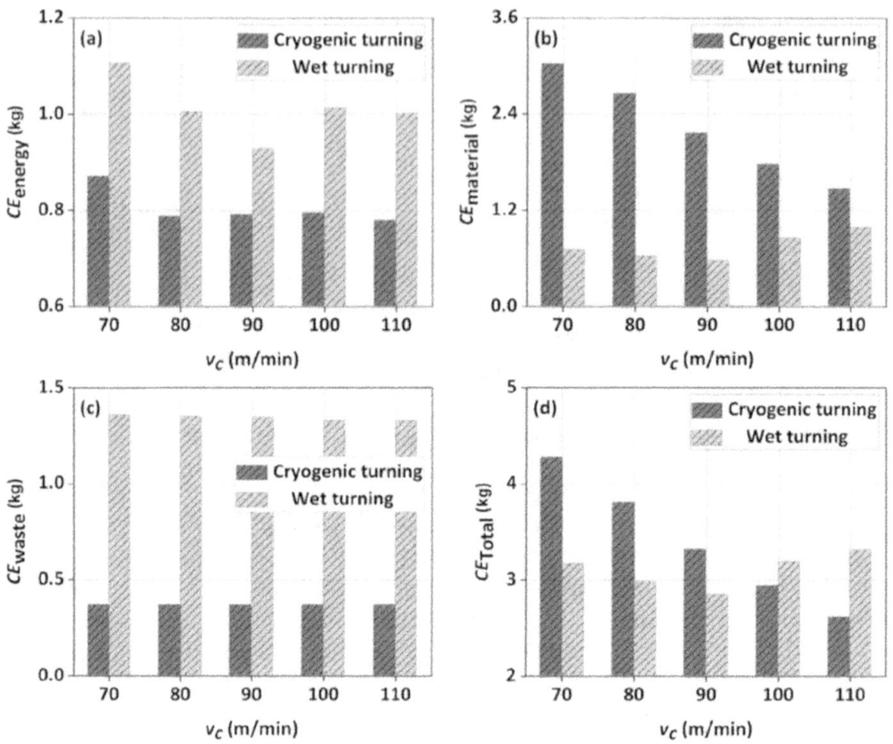

FIGURE 1.25 Comparison of carbon emissions for turning of Ti-6Al-4V under wet and cryogenic environments. (From Ref. [79].)

60.60% for dry, EMQL, and LCO_2 cutting environments, respectively, for machining of AMed IN625 in comparison to the wrought alloy, as shown in Figure 1.27 [18]. Tool life increased during the machining of AMed IN625 by 191.55%, 115.80% and 65.95% for dry, EMQL and LCO_2 environments, respectively, as compared to the wrought IN625 alloy, as observed in Figure 1.28 [18].

Nandam et al. [80] investigated the machinability of tungsten heavy alloys using carbide cutting tools through cryogenic and conventional coolants in terms of material removal rate, surface integrity and cutting forces. Superior machining performance has been noticed for the cryogenic environment, where material removal rate is three times higher than conventional machining. Surface roughness and cutting forces are substantially reduced due to the cryogenic environment. Jamil et al. [81] investigated milling of titanium alloy using different sustainability measures (specific cutting energy, carbon emissions, energy efficiency and process time) and machinability under cooling and lubrication techniques such as MQL, CO_2-snow, cryogenic LN_2 and dry cutting. From the experimental observations, it is evident that CO_2-snow outperformed cryogenic LN_2, MQL and dry cutting, as shown in Figure 1.29 [81], Figure 1.30 [81] and Figure 1.31 [81].

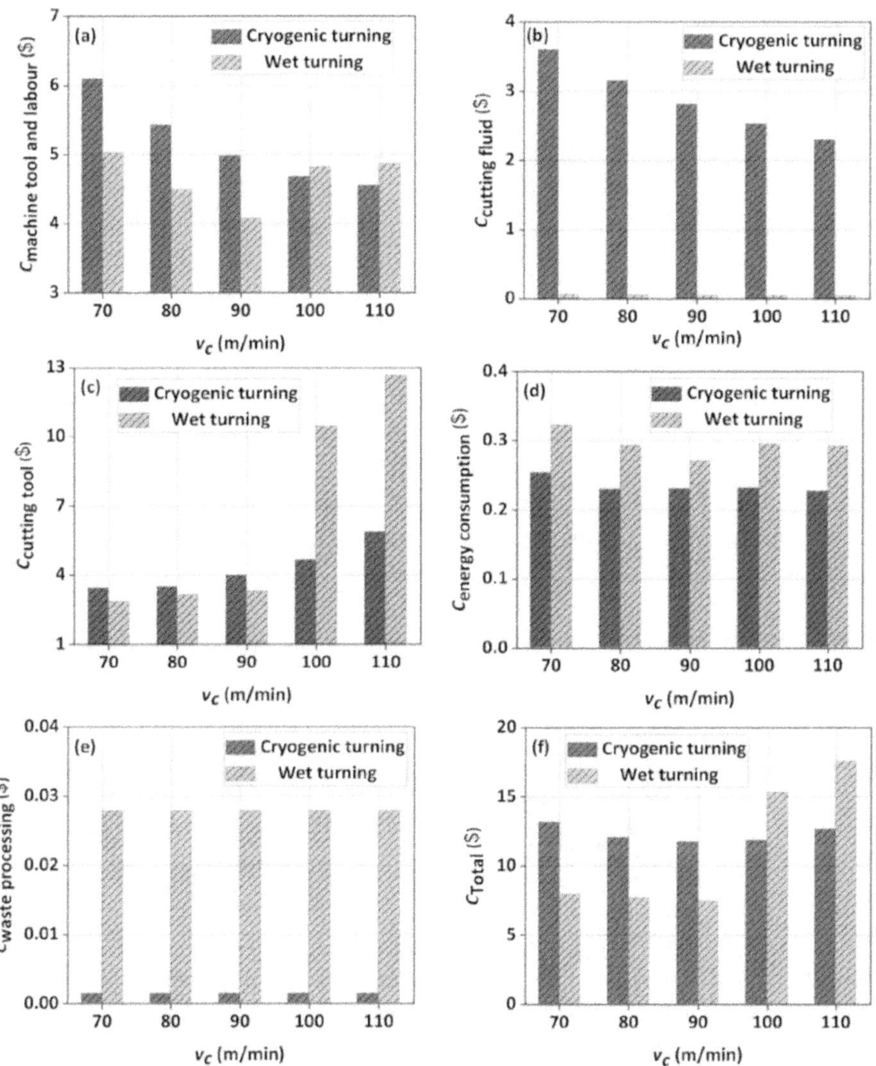

FIGURE 1.26 Comparison of cost components for turning of Ti-6Al-4V under cryogenic and wet environments. (From Ref. [79].)

1.3.4 Air Cooling

Application of chilled air during machining to cool the workpiece and tool has now become a new technique, and some researchers have experimented with it [82–88]. Air cooling is considered the most environmentally friendly and cleanest machining condition and enhances the tool life. It is evident that the surface roughness increases with the application of air cooling in machining as compared to dry cutting. But the surface roughness obtained during air cooling is higher than that during MQL.

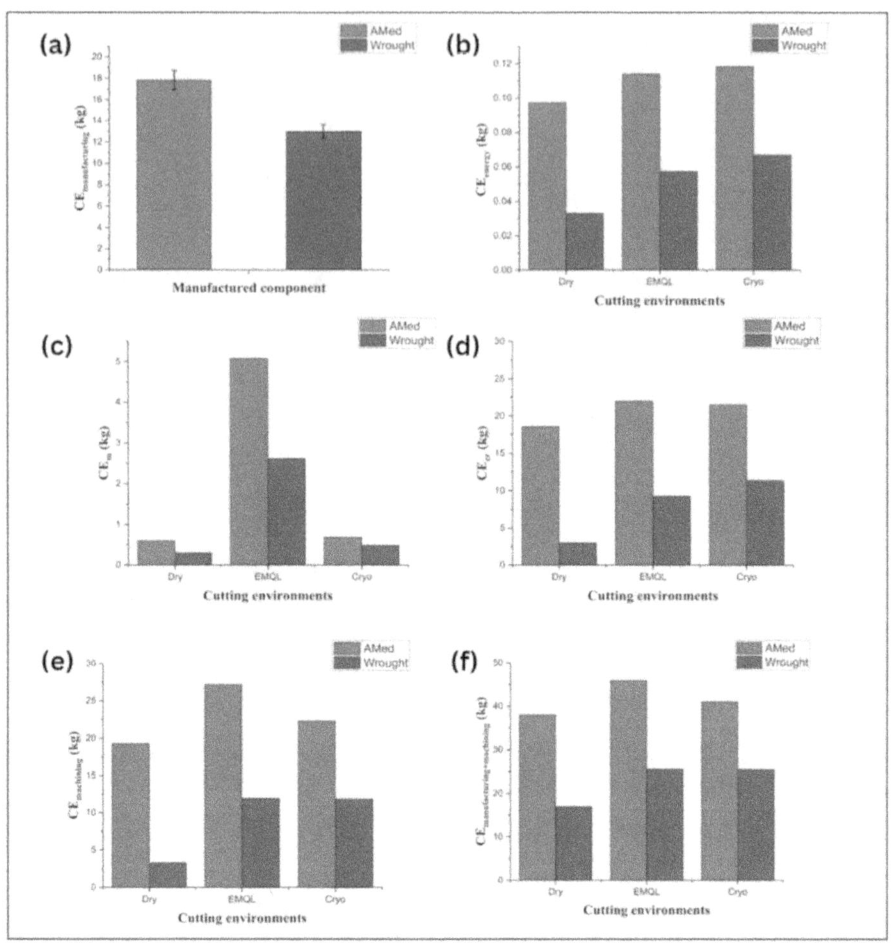

FIGURE 1.27 Comparison of CE from fabrication as well as machining of AMed and wrought IN625 under dry, EMQL and LCO_2 cutting environments: (a) CE from the fabrication stage of AMed and wrought IN625, (b) CE from the energy consumption, (c) CE from cutting tool utilization, (d) CE from waste disposal, (e) CE from chip recycling and (f) CE from the machining stage. (From Ref. [18].)

The effect of chilled air on the surface finish is highly dependent on the machining parameters. In general, it could be claimed that air cooling produces lower surface roughness than dry cutting. However, the produced surface roughness is higher than that made by MQL or emulsion coolant [89–91]. Liu and Kevinchou [92] investigated turning of A390 aluminium with uncoated carbide insert. The reduction of flank wear has been found to be 20% using air cooling with cutting speed of 5 m/s and feed rate of 0.055 mm/rev as parametric conditions. It also reduces the chip–tool contact temperature by 7%. The cutting force has been reduced because of the reduction of the

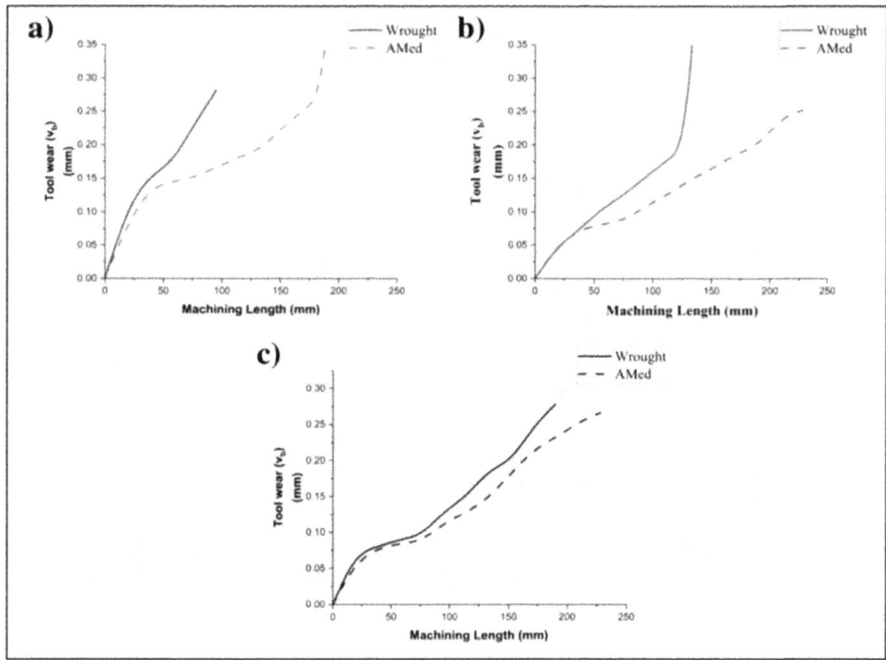

FIGURE 1.28 Comparison of tool wear with respect to machining length of AMed and wrought IN625 for different cutting environments: (a) dry, (b) EMQL and (c) LCO_2. (From Ref. [18].)

adhesion mechanism and BUE formation. The suitability of the air cooling process is completely dependent on the machining parameters.

1.3.5 Solid Lubricants

Application of solid lubricants in machining is gaining importance nowadays because of their lubricating properties, lower toxicity and biodegradability. Common solid lubricants used in machining by several researchers are molybdenum disulphide (MoS_2), calcium fluoride (CaF_2), graphite and boric acid (H_3BO_3) [93, 94]. Reddy and Rao [95] conducted machining of AISI 1045 steel with different tool geometries of cutting inserts using graphite and MoS_2 solid lubricants to study cutting force, surface roughness and specific energy. MoS_2 solid lubricant assisted machining performed well as compared to graphite based application due to significant reduction of friction at the rubbing surfaces. Shaji and Radhakrishnan [96] applied graphite solid lubricants in surface grinding to lower the cutting temperature. Significant reduction of cutting force, temperature, specific energy and surface roughness have been noticed using solid lubricants as compared to conventional coolant. Application of graphite solid lubricant in milling of AISI 1045 steel has been investigated to lower the heat generation [97]. The milling performance has been improved a lot with respect to cutting

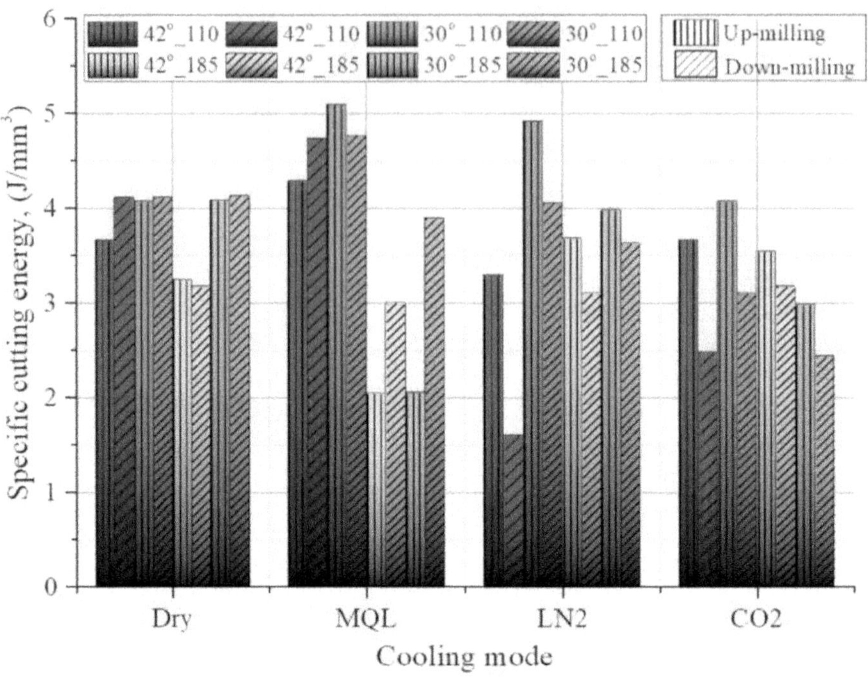

FIGURE 1.29 Specific cutting energy at varying cutting speed, helix angle and cutting environments. (From Ref. [81].)

force, specific energy requirements and surface roughness. Singh and Rao [98] investigated the performance of hard turning of bearing steel with mixed ceramic cutting tools using graphite and MoS_2 solid lubricants. Surface roughness has been reduced through solid lubricant assisted hard turning as compared to dry hard turning and graphite solid lubricant, as is evident from Figure 1.32 [98].

Mukhopadhyay et al. [99] conducted turning of AISI 1040 steel using solid lubricants with coated carbide inserts, and improvement of performance has been noticed. But the application of solid lubricants in machining is found to be more difficult than conventional cutting fluids.

1.3.6 Spray Impingement Cooling

Spray impingement cooling refers to the application of cutting fluid in the machining zone at high speed (350–500 km/h) and at high pressure (5.5–35 MPa) [100]. There is evidence of effective impingement of coolant in the machining zone through spray or high pressure cooling and also superior chip-breaking capabilities. Mia et al. [101] observed that there is superior heat dissipation through spray cooling application machining of Ti-6Al-4V. There is an issue related to tool wear through spray cooling capability [102]. Kumar et al. [103] conducted machining of hardened AISI D2 steel under environmentally conscious conditions such as spray impingement cooling to investigate machinability characteristics by measuring tool wear, cutting

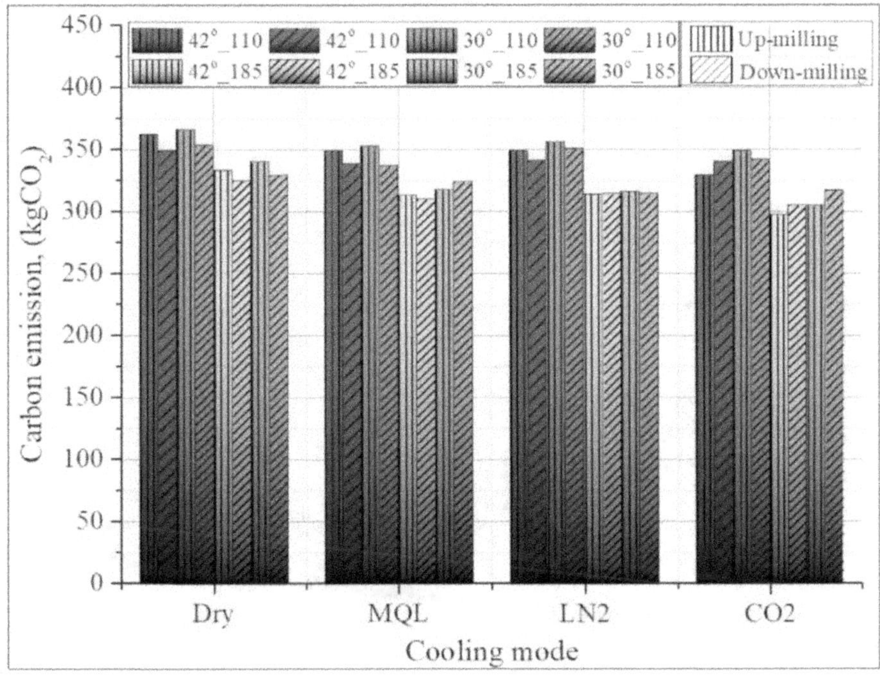

FIGURE 1.30 Carbon emissions (kg-CO_2) against milling parameters and coolant modes. (From Ref. [18].)

temperature and surface roughness. Flank wear and surface roughness are well within the recommended limits of 0.2 mm and 1.6 μm respectively in all the runs, which shows the benefit of application of air-water spray cooling in machining. The images of flank wear at respective experimental runs as per the Taguchi L_{16} orthogonal array design of the experiment are shown in Figure 1.33 [103]. The proposed spray cooling technique reduces the cutting temperature because the latent heat is being absorbed by the water droplets through evaporation, and thus it reduces the growth of tool wear and improves the surface quality and tool life. Cutting temperature lies between 120.4°C and 217°C. The images of cutting temperatures for runs 1–4 are shown in Figure 1.34 [103]. From an industrial application point of view, it may be considered as a stable, environmentally friendly and ecologically cleaner method of machining hardened AISI D2 steel using multilayer coated carbide inserts.

Kumar et al. [104], in another investigation on machinability of AISI D2 steel, applied spray cooling to study the chip reduction coefficient and flank wear and developed ANN prediction models using a feed-forward back-propagation algorithm. The model is found to be accurate and effective in predicting the responses as the R^2 value is high with minimum absolute error. Mishra et al. [105] carried out machining of AA 7075/SiC particulate MMCs under dry and spray cooling environments and optimized the machining parameters through a multi-objective optimization technique called grey relational analysis combined with the Taguchi method. Water spray cooling has been proved to be beneficial compared to dry cutting

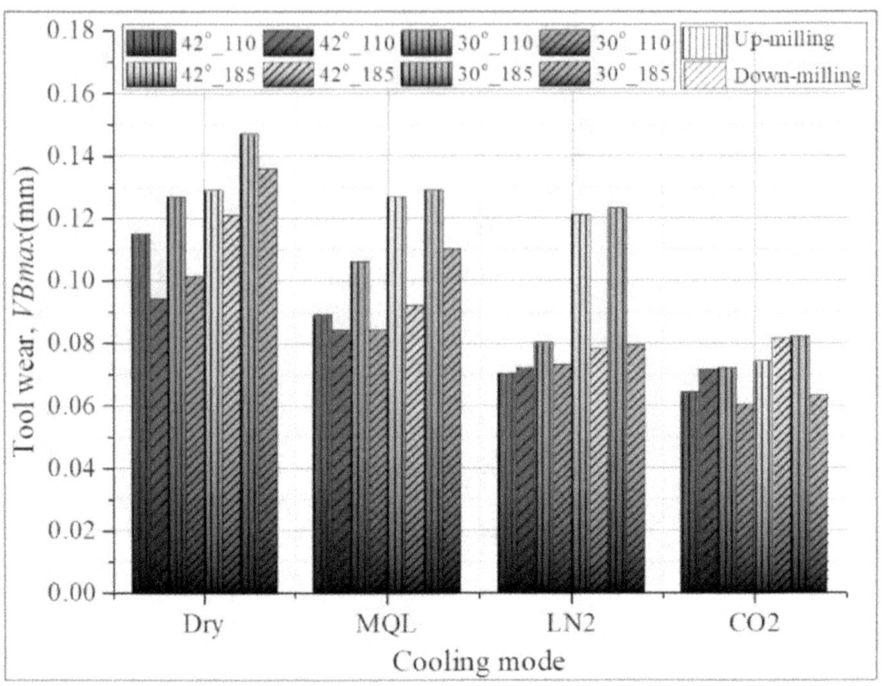

FIGURE 1.31 Tool wear result against milling parameters under sustainable cooling modes. (From Ref. [18].)

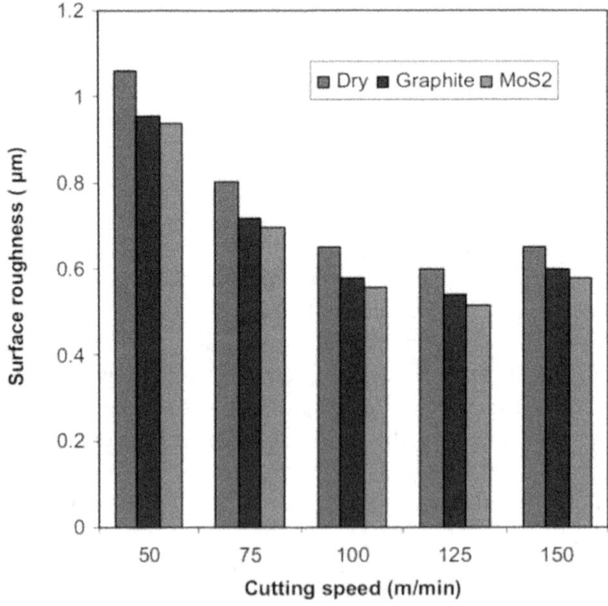

FIGURE 1.32 Bar graph showing surface quality improvement. (From Ref. [98].)

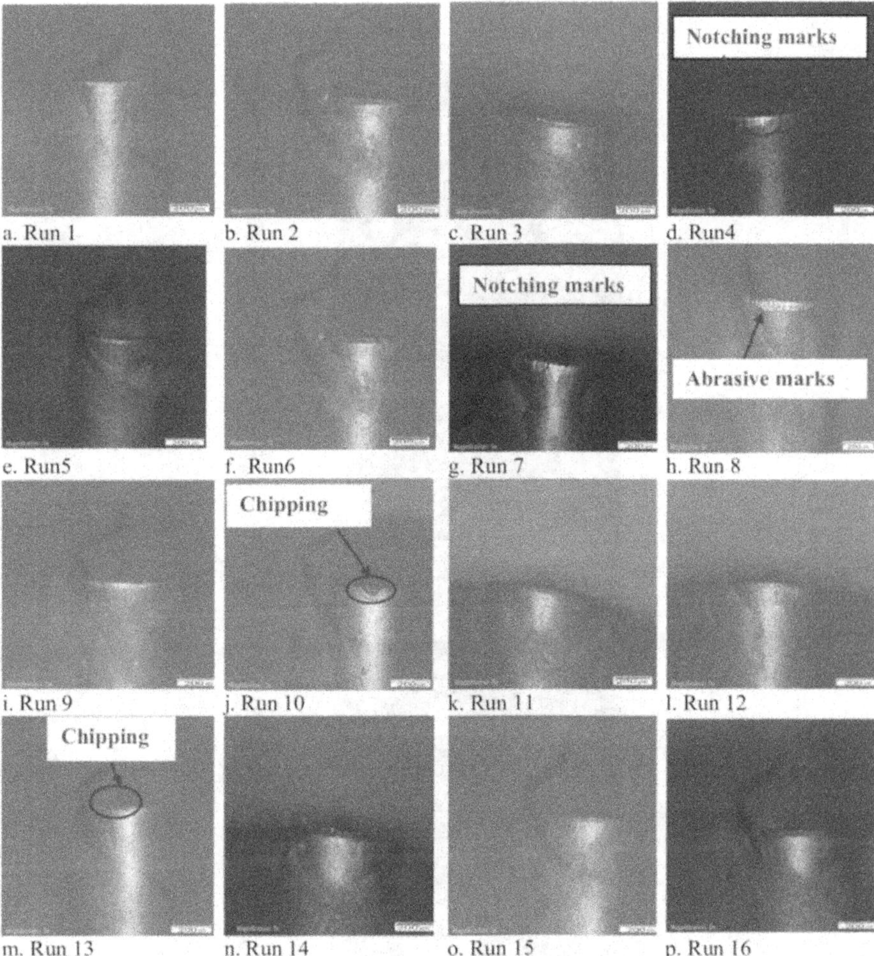

FIGURE 1.33 (a–p) Flank wear images of respective experimental runs. (From Ref. [103].)

based on machinability studies of composite materials and has also been shown to be environmentally benign. Sahu et al. [106] studied the comparative machining performance of AISI 1045 steel under dry and spray cooling environments by measuring material removal rate (productivity), cutting temperature and surface roughness. The machinability performance of environmentally conscious spray cooling is improved compared to dry machining. The optimized results yield maximization of material removal rate with minimum surface roughness and temperature. Ukamanal et al. [107] developed TiO_2 nanoparticles from a high energy ball milling process and applied them in machining of AISI 316 stainless steel using water based spray cooling. From experiment, it is observed that at 0.03% of weight concentration of nanofluid, tool and chip temperatures were found to be lower. Ukamanal et al. [108] optimized machining and spray cooling parameters for multi-responses

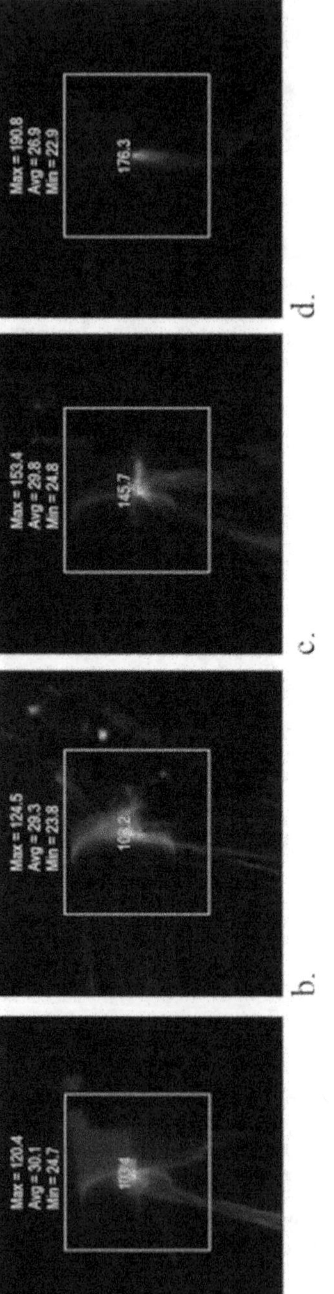

FIGURE 1.34 Image of cutting temperature using an infrared camera at (a) run 1, (b) run 2, (c) run 3 and (d) run 4. (From Ref. [103].)

(tool temperature, chip temperature, surface roughness and flank wear) during the machining of AISI 316 stainless steel through WPCA coupled with the Taguchi method. Machining under spray cooling is found to be effective compared to dry machining, as substantial reduction of cutting temperature is noticed. Kumar et al. [109] investigated the machinability of hardened AISI D2 steel using a coated carbide insert through air assisted nanofluid impingement spray cooling (Al_2O_3-water and TiO_2-water nanofluids). The most favourable results are noticed at 0.01% weight concentration of TiO_2, and at this condition, compared with the same concentration of Al_2O_3 nanofluid, 29% reduction in tool-flank wear, 9.7% drop in cutting temperature and 14.3% reduction in surface roughness are found, as shown in Figure 1.35 and Figure 1.36 respectively [109]. Keeping an eye on the outperformance of TiO_2

Types of Nano fluids	Test run-1 VBc	Test run-2 VBc	Test run-3 VBc	Average VBc
Al_2O_3 0.005 % weight concentration	Micro-groove VBc = 0.067 mm 200 μm	VBc = 0.058 mm 200 μm	VBc = 0.076 mm 200 μm	0.067 mm
TiO_2 0.005 % weight concentration	Abrasion VBc = 0.066 mm 200 μm	VBc = 0.057 mm 200 μm	VBc = 0.063 mm 200 μm	0.062 mm
Al_2O_3 0.01 % weight concentration	Micro-grooves VBc = 0.058 mm 200 μm	Micro-groove VBc = 0.055 mm 200 μm	VBc = 0.052 mm 200 μm	0.055 mm
TiO_2 0.01 % weight concentration	VBc = 0.038 mm 200 μm	Micro-groove VBc = 0.036 mm 200 μm	VBc = 0.043 mm 200 μm	0.039 mm
Al_2O_3 0.05 % weight concentration	VBc = 0.092 mm 200 μm	VBc = 0.097 mm 200 μm	VBc = 0.114 mm 200 μm	0.101 mm
TiO_2 0.05 % weight concentration	VBc = 0.083 mm 200 μm	VBc = 0.095 mm 200 μm	VBc = 0.101mm 200 μm	0.093 mm

FIGURE 1.35 Experimental results of nose flank wear. (From Ref. [109].)

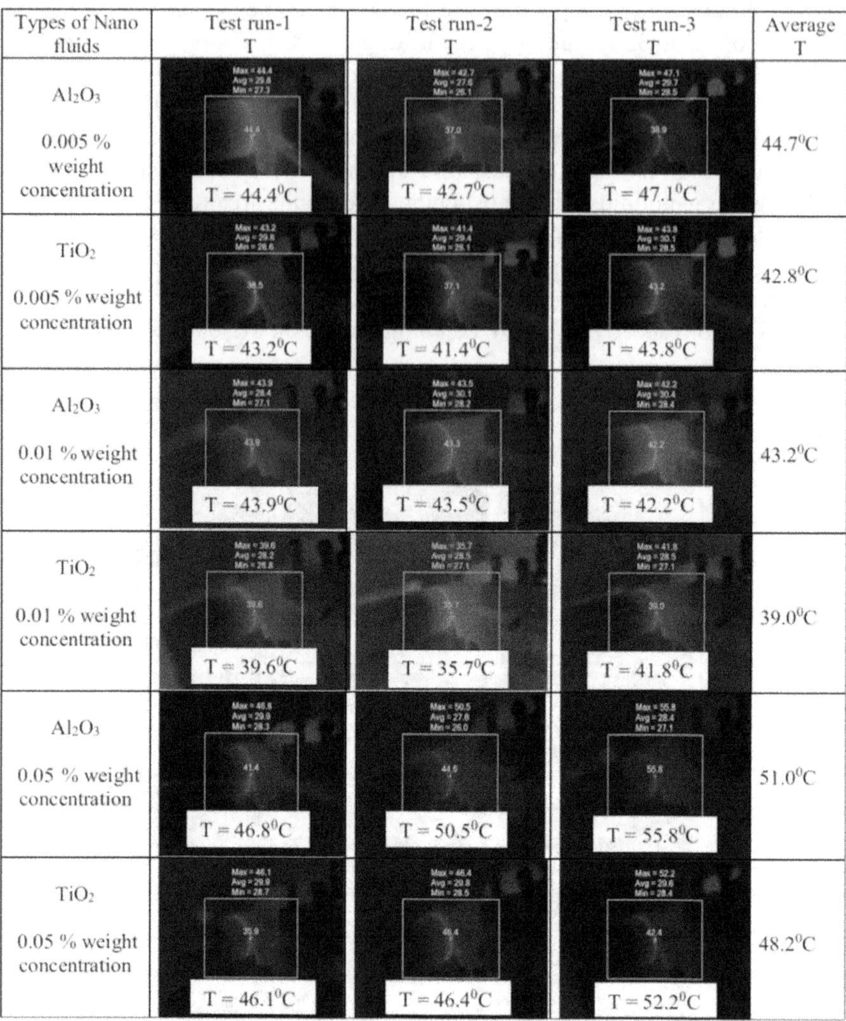

Types of Nano fluids	Test run-1 T	Test run-2 T	Test run-3 T	Average T
Al_2O_3 0.005 % weight concentration	T = 44.4°C	T = 42.7°C	T = 47.1°C	44.7°C
TiO_2 0.005 % weight concentration	T = 43.2°C	T = 41.4°C	T = 43.8°C	42.8°C
Al_2O_3 0.01 % weight concentration	T = 43.9°C	T = 43.5°C	T = 42.2°C	43.2°C
TiO_2 0.01 % weight concentration	T = 39.6°C	T = 35.7°C	T = 41.8°C	39.0°C
Al_2O_3 0.05 % weight concentration	T = 46.8°C	T = 50.5°C	T = 55.8°C	51.0°C
TiO_2 0.05 % weight concentration	T = 46.1°C	T = 46.4°C	T = 52.2°C	48.2°C

FIGURE 1.36 Experimental results of cutting temperature. (From Ref. [109].)

nanofluids (0.01 wt%) in hard machining, this may be adopted in other machining processes through spray cooling.

1.3.7 Hybrid Cooling

Recently, hybrid cooling and lubrication strategies have been adopted to gain advantages in sustainable manufacturing. The use of cryogenic MQL is one of the examples that have shown improvements in machinability performance during turning, drilling, milling and grinding etc., where gases are liquefied at or below –150°C and passed through a specially designed nozzle [110]. Su et al. [111] observed the improvement of tool life by 124% through cryogenic MQL machining of Inconel 718 alloy, and various researchers have also obtained significant improvement [112–114]. Sartori et al. [115] used hybrid cooling technologies like MQL with liquid N_2 or CO_2 in machining. It is evident from the observations that crater wear has been substantially reduced with an increase of surface quality. Pusavec et al. [116, 117] applied hybrid cooling techniques such as MQL with cryogenic cooling for sustainable manufacturing of Inconel materials, and modelling and optimization of the process have been done for sustainability. There is a significant improvement of the results noticed during the investigation through the hybrid system. Gupta et al. [118] investigated the effectiveness of different cooling and lubrication techniques such as dry, liquid nitrogen and hybrid liquid nitrogen with MQL (LN_2+MQL) environments during the machining of titanium alloy. Various responses during machining have been recorded to investigate the machinability, such as surface roughness, cutting force and temperature, as well as environmental parameter impact (cycle time, productivity, energy consumption, carbon emission) and economic analysis. With the help of an analytic hierarchy process (AHP) coupled with the technique for order preference based on similarity to ideal solution (TOPSIS), sustainability assessment has been carried out. There is evidence of excellent cooling and lubrication impacts of hybrid liquid nitrogen with MQL conditions during machining. With respect to dry turning, the enhancement in cycle time and productivity of LN_2 and LN_2+MQL was observed to be 29.01% and 34.21% respectively, as shown in Figure 1.37 [118]. Investigation of energy consumption under dry, LN_2 and LN_2+MQL conditions at different machining parameters, such as cutting speed: 100, 125 and 150 m/min and feed rate: 0.1, 0.125 and 0.15 mm/rev, has been done for titanium alloy. Reduction in energy consumption has been achieved with lowering of cutting speed and feed. There is an average reduction in energy consumption of up to 25.05%, 23.55% and 23.01% at the lowest cutting speed in machining under dry, LN_2 and LN_2+MQL conditions, respectively, as shown in Figure 1.38 [118]. The best outcomes for the overall sustainability index have been achieved through lower cutting parameters under a hybrid LN_2+MQL environment, as shown in Figure 1.39 [118].

1.4 ADVANCES IN TEXTURED CUTTING TOOLS

The evolution of specially designed micro-/nano-surface textured cutting tools is gaining importance for sustainable hard machining nowadays for better machinability. It offers many advantages for machinability improvements, such as higher tool life with better surface quality, reduction of tool wear, cutting temperature and cutting force etc. This is because textured cutting tools behave like a reservoir for coolant and consequently minimize the chip–tool interface friction and wear rate. This is the new process of machinability performance improvement through application of coolants by specially designed textured grooves on the cutting tool surface. Recently, some researchers have investigated the influence of textured cutting tools during hard machining towards sustainability. The cutting tool surfaces, such as both

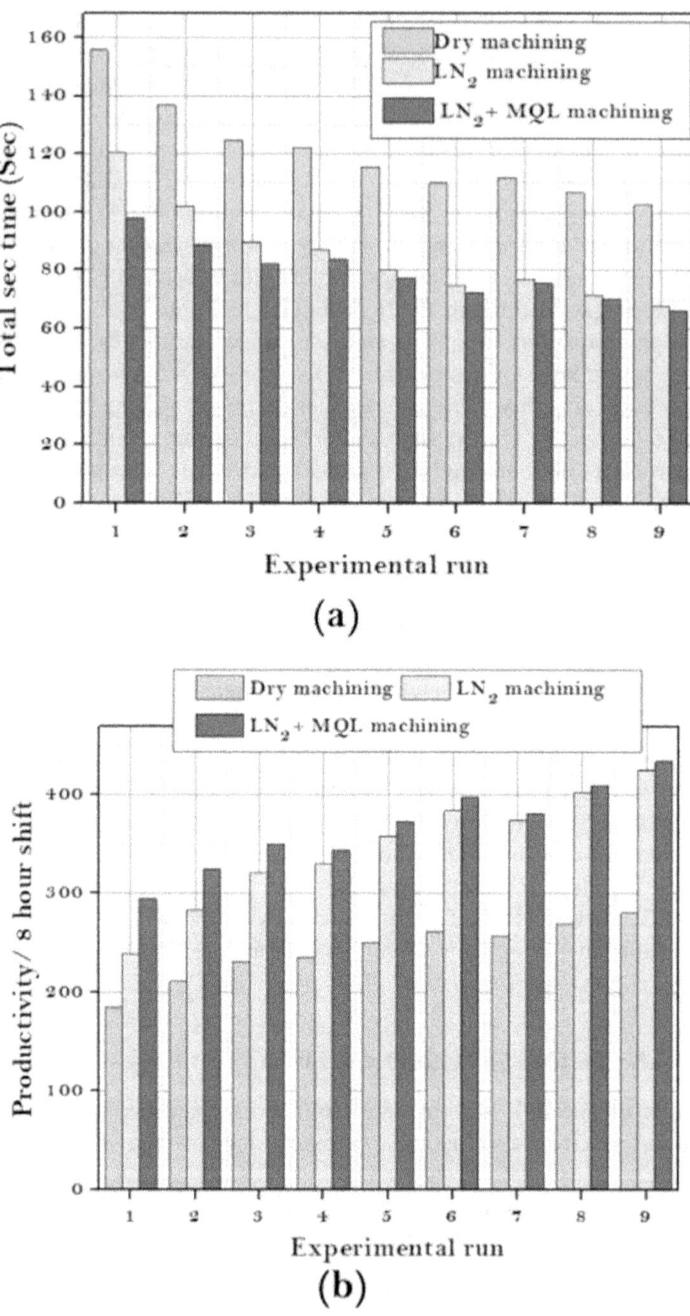

FIGURE 1.37 (a) Total cycle time and (b) productivity with respect to each experimental run. (From Ref. [118].)

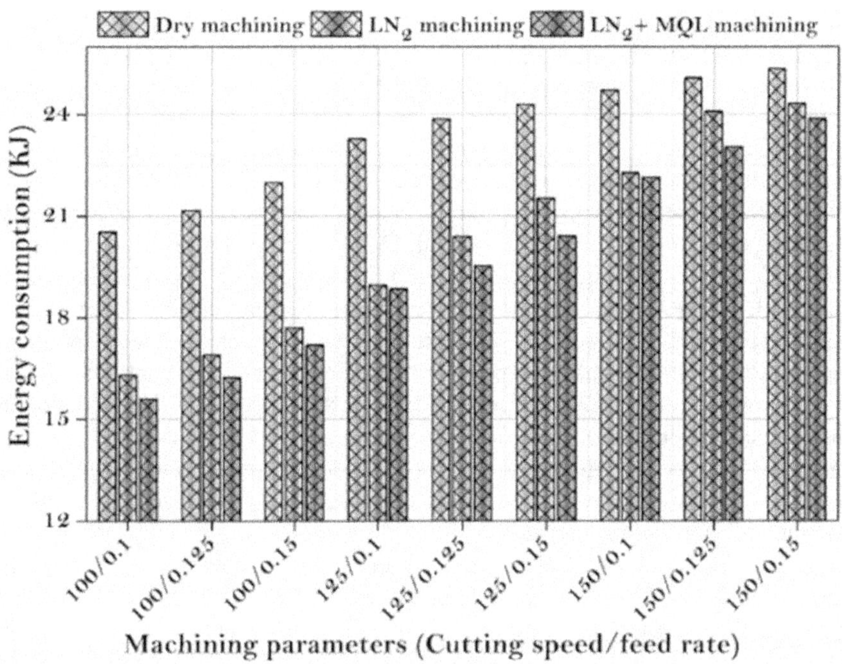

FIGURE 1.38 Energy consumption values under dry, LN_2 and LN_2+MQL environments at different experimental runs. (From Ref. [118].)

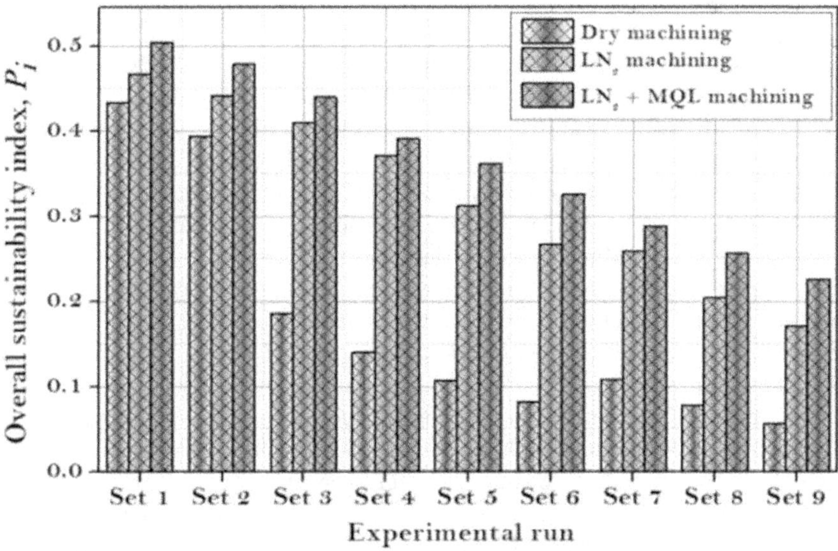

FIGURE 1.39 Overall sustainability index and sustainability assessment results under dry, LN_2 and LN_2+MQL machining. (From Ref. [118].)

flank and rake surface, are textured specially for the effectiveness of cooling and lubrication strategies in hard machining.

There are various techniques adopted for surface grooving/texturing, namely laser surface texturing (LST), focused ion beam (FIB), electro-discharge machining (EDM) and cathodic arc vacuum evaporation etc. A popular method adopted for texturing at both flank and rake surface of the cutting tool is LST. Various types of textured patterns are also grooved on the surfaces, such as perpendicular, parallel type and crossed pattern etc. [119]. Koshy et al. [120] developed a textured cutting tool through EDM and compared the machining performance with that of a non-textured cutting tool. It is evident from the turning experiment that cutting force and feed forces are reduced by 13% and 30% respectively as compared to untextured cutting inserts, due to self-lubrication properties. Yunsong et al. [121] fabricated nano-textured carbide tools using femtosecond LST, further coated by WS_2 soft coatings using magnetron sputtering, multi-arc ion plating and ion beam assisted deposition techniques. The machinability has been improved as compared to untextured cutting tools because the responses such as cutting temperature, cutting force and coefficient of friction are substantially reduced.

Sugihara and Enomoto [122] developed nano-/micro-grooves on the rake surface of cutting tools through femtosecond laser technology and studied the machinability. Decrease of the adhesion of aluminium chips is observed under wet machining operation as compared to dry. The diamond-like carbon (DLC) coated cutting tool with banded nano-/micro-textured surface consisting of nano-/micro-textured and mirror-polished areas (as shown in Figure 1.40) enhanced anti-adhesive effects and lubricity of the cutting tool surface, imparting a low coefficient of friction (Figure 1.41).

Wenlong et al. [123] developed micro-holes on the flank and rake face of the carbide cutting tool through a micro-EDM method. The machinability performance of cutting inserts has been observed to be better by providing minimum tool wear. Devaraj et al. [124] studied the machinability of aluminium MMCs using textured cutting tools and looked at the effect of micro-hole design parameters on machining performance. Micro-hole textured cutting inserts behave like chip breakers with solid lubricant incorporation and satisfy both cooling and lubrication functions. Therefore, this reduces surface roughness, power consumption and tool wear during machining. Sharma et al. [125] fabricated micro-textured grooves on coated carbide inserts using a femtosecond laser machine. A spiral triangular texture behaves like a lubricant reservoir and provides a larger area for heat dissipation. The surface coatings of carbide inserts are not eradicated by using a femtosecond laser machine, which enables the machining of hard materials. Figure 1.42 and Figure 1.43 show the design of a spiral triangular texture with dimensions of the actual texture grooving, respectively.

Sugihara and Enomoto [126] used femtosecond laser technology to fabricate micro-stripe textured cutting tools on both the flank and rake surface and studied their performance in milling operation. Flank wear has been lowered significantly during face milling of steel materials, which offers excellent wear resistance. The crater wear has been suppressed during machining of medium carbon steel by micro-stripe textured cutting tools and acts as a reservoir for cutting fluids. Severe crater wear of 9 μm deep and 200 μm wide has been observed in conventional tools as

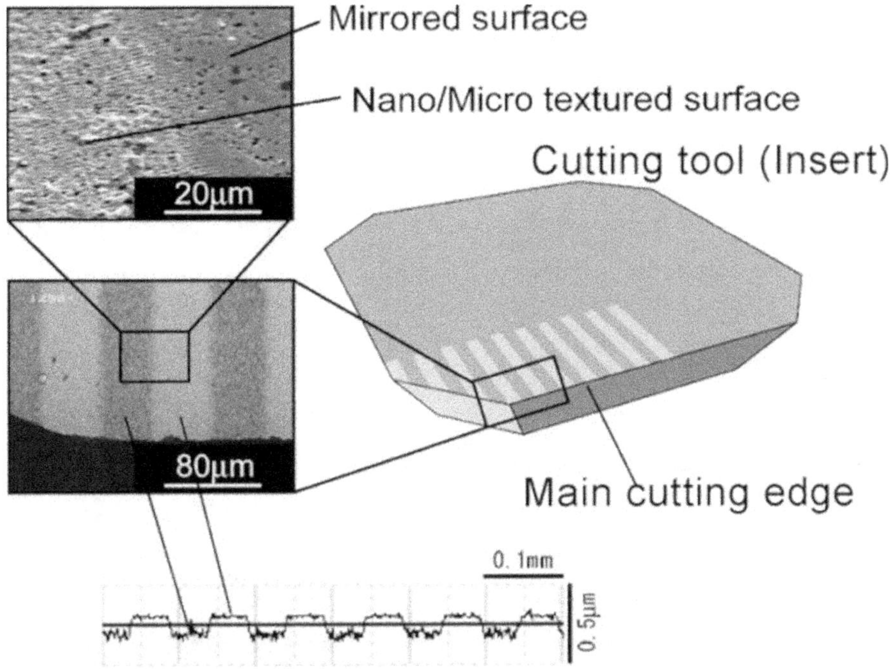

FIGURE 1.40 Newly developed cutting tool with banded nano-/micro-textured surface. (From Ref. [122].)

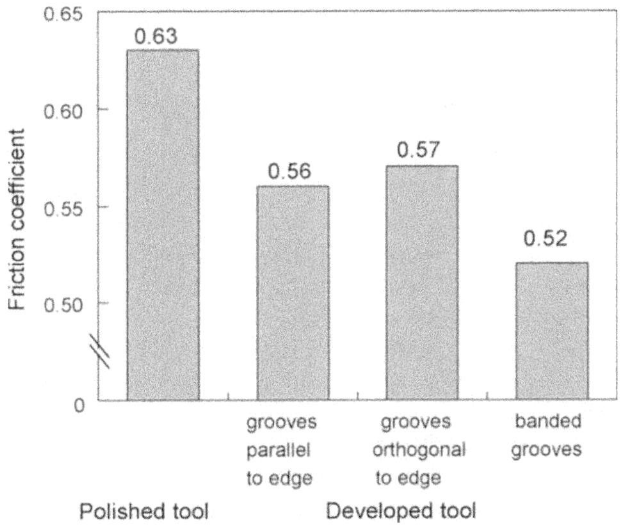

FIGURE 1.41 Coefficient of friction on rake face. (From Ref. [122].)

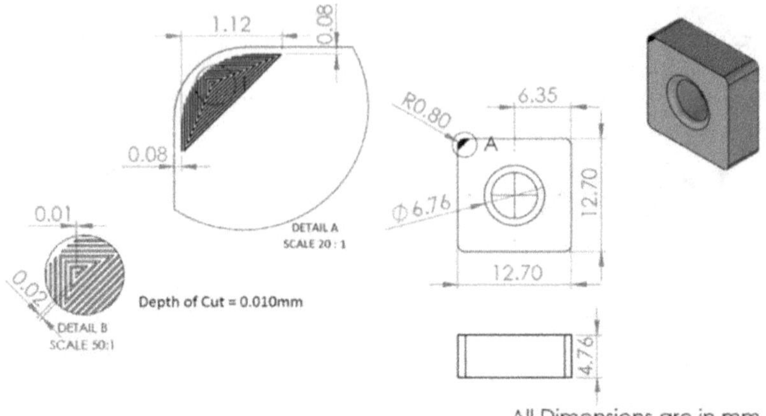

FIGURE 1.42 Design of micro-texture with dimensional detail. (From Ref. [125].)

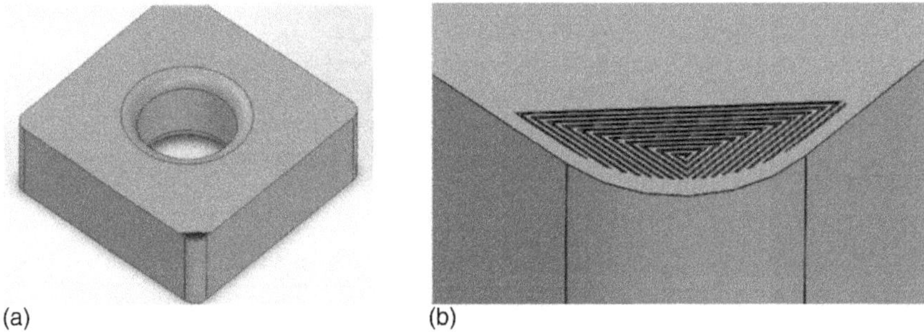

FIGURE 1.43 (a) Three-dimensional representation of location and orientation micro-texture design and (b) spiral triangular micro-texture design on the rake face of cutting insert. (From Ref. [125].)

shown in Figure 1.44 (a). The crater wear has been significantly reduced from 9 μm to 3 μm for micro-stripe grooves parallel to the edge, shown in Figure 1.44 (b), while there is a slight suppression of wear for micro-stripe grooves orthogonal to the main cutting edge, as observed in Figure 1.44 (c).

Ranjan et al. [127] observed that there is a significant improvement of machinability performance using a textured cutting tool as compared to a conventional tool in terms of tool wear, surface roughness, cutting force, cutting temperature and friction. Xie et al. [128] observed that there is a reduction of cutting force of 32.7% utilizing micro-grooved textured tools over non-textured cutting tools during machining. On decreasing micro-groove depth from 149 μm to 25 μm, cutting temperature decreases and shearing angle increases. However, cutting temperature greatly increases and shearing angle greatly decreases at a micro-groove depth of 7 μm and aspect ratio of

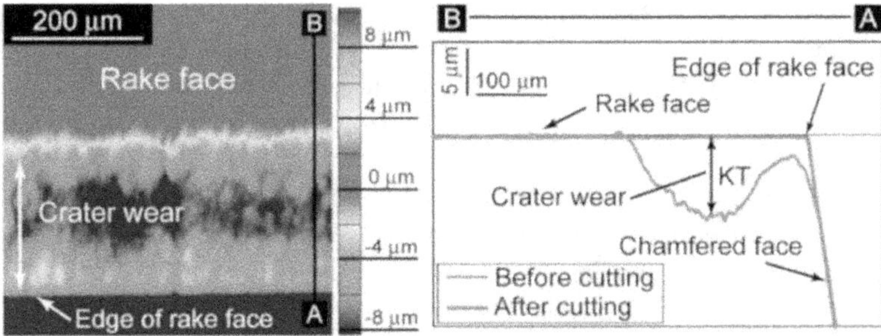

(a) Conventional tool

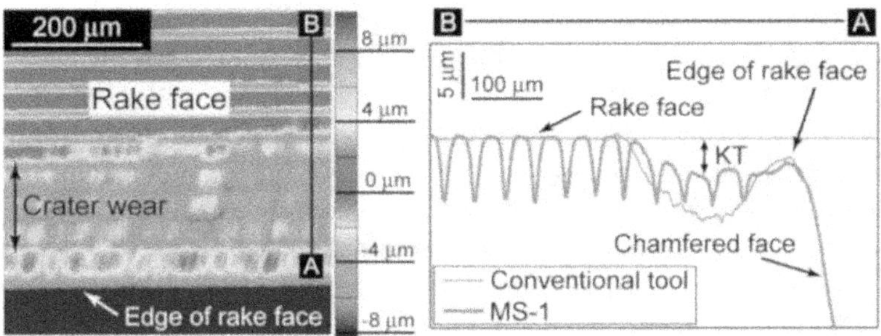

(b) MS-1 (5 μm deep grooves parallel to edge)

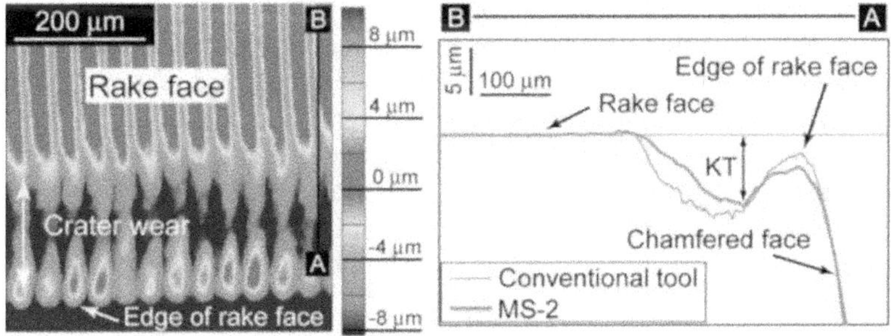

(c) MS-2 (5 μm deep grooves orthogonal to edge)

FIGURE 1.44 Profiles of the rake face of cutting tools after wet cutting for 300 m (left: three-dimensional profile; right: sectional profile). (a) Conventional tool, (b) MS-1 (5 μm deep grooves parallel to edge) and (c) MS-2 (5 μm deep grooves orthogonal to edge). (From Ref. [126].)

0.14. This is because the aspect ratio is too little to form air spaces between cutting chip back and tool rake surface. The micro-grooved tool may decrease cutting temperature by 103°C and more compared with a traditional plane tool. Figure 1.45 shows cutting temperature curves for the micro-grooved tool and traditional plane tool. Figure 1.46 shows the scanning electron microscope (SEM) micro-topographies of cutting chips for the micro-grooved tool and traditional plane tool [128]. It is seen that the micro-grooved tool produced a finer saw-tooth on the chip free surface than the traditional plane tool, leading to a more stable cutting state.

Kawasegi et al. [129] and Sasi et al. [130] investigated the machinability performance using a textured cutting tool for cutting force and its effects. The performance of nano-textured cutting tools has been found to be better as compared to micro-textured tools. With increase of cutting speed, cutting force decreases because of lower adherability of the work material or BUE. Fatima et al. [131] developed a micro-grooved texture on both the flank and rake surface of an uncoated carbide insert and studied the machinability of AISI 4140 steel. It is evident that cutting force and feed force decrease 10% and 23% respectively because of less chip–tool interface contact. Sun et al. [132] fabricated micro-grooves, micro-pits and hybrid texture combining micro-pits and micro-grooves through a laser technique on the rake face of carbide tools. The machining performance of pure iron using MoS_2 solid lubricant has been investigated using a hybrid-textured cutting tool, single-textured cutting tool and conventional tool. Figure 1.47 shows different micrographs of surface textures on the rake face of a carbide insert, where Figure 1.47 (d) indicates the micrograph of a hybrid-textured self-lubricating tool, Figure 1.47 (e) shows the cross-section micrograph of a micro-groove and Figure 1.47 (f) shows the cross-section micrograph of a micro-pit. The machinability performance of hybrid-textured cutting tools has been observed to be better in terms of cutting forces, cutting temperature, friction coefficient at the tool–chip interface, shear angle, surface roughness of the machined workpiece, chip morphology and tool wear on the rake face. Hybrid-textured cutting tools acts like a micro-reservoir for constant lubricant replenishment of micro-grooves and micro-pits and are suitable for dry machining due to self-lubrication. At a higher cutting speed of 120 m/min, the effect of hybrid texture has been found to be beneficial where the reductions of cutting forces were 7.1–33.3% and 6.9–21.7% with cutting temperatures as shown in Figure 1.48 [132]. The reduction of the surface roughness has been found to be 42.9–69.1%.

Chang et al. [133] conducted milling operation through a micro-groove cutter perpendicular to the cutting edge over other textured cutting tools. The observed flank wear is lower using a micro-groove cutter, and this may be attributed to the lower adhesion of work and tool materials. Kummel et al. [134] concluded that the machining performance of a dimple-textured carbide insert outperformed a channel textured cutting tool due to lower tool wear, and this may be due to the increased adhesive properties as well as better mechanical interlocking. Liu et al. [135] fabricated textured grooves parallel, perpendicular and inclined to the main cutting edge at the flank surface by fibre laser, as shown in Figure 1.49, and noticed that the parallel grooves (AT-1) induced higher wear resistance during machining of green alumina ceramic. There is severe abrasive wear at the flank surface while the rake surface

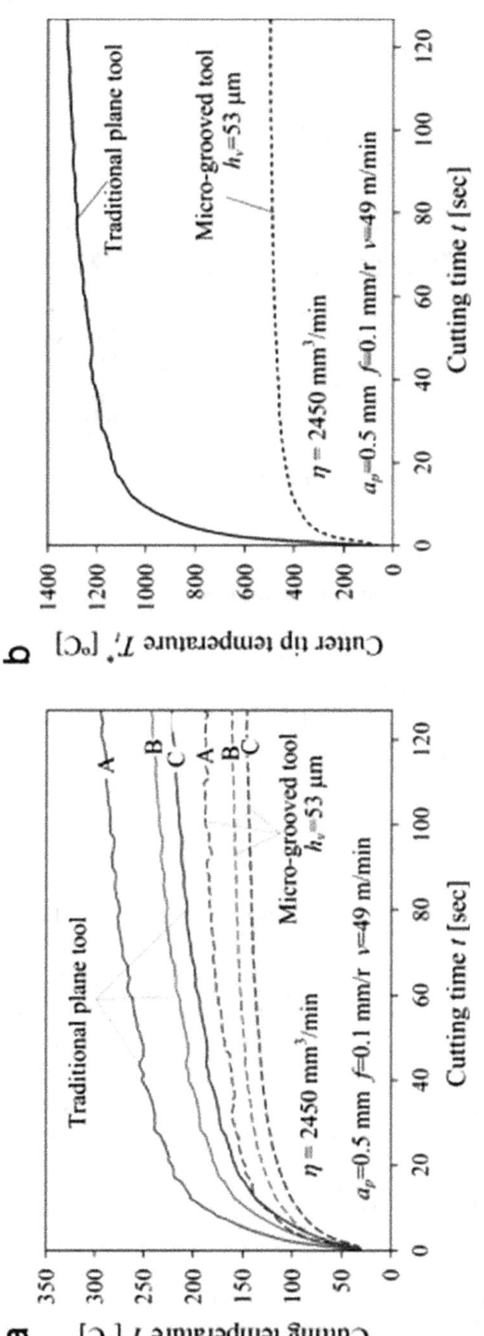

FIGURE 1.45 Cutting temperatures for traditional plate tool and micro-grooved tool: (a) the cutting temperature T on tool rake surface and (b) the predicted cutter tip temperature Tt. (From Ref. [153].)

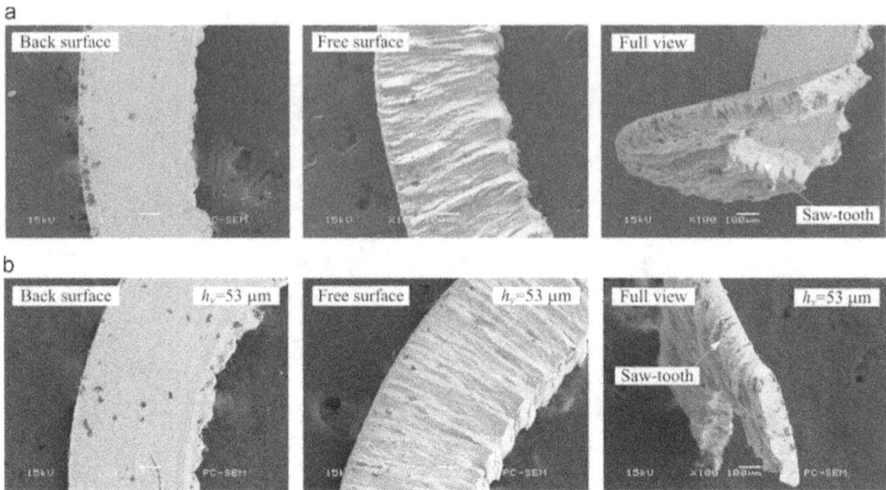

FIGURE 1.46 Cutting chip micro-topographies: (a) traditional plane tool and (b) micro-grooved tool. (From Ref. [153].)

undergoes minimal wear. Textured cutting tools have been observed to be more wear resistant than untextured carbide tools, whereas the AT-1 tool proved to have the highest wear resistance as compared to other inserts, as shown in Figure 1.50 [135].

Arulkirubakaran et al. [136] investigated the machining performance of Ti-6Al-4V alloy by untextured and textured coated and uncoated carbide cutting tools. There is a reduction of flank wear by 29.4% and higher wear resistance was observed using textured a TiAlN coated cutting tool perpendicular to the direction of flow of chip. Thomas et al. [137] investigated machinability studies of mild steel and aluminium alloy using various textured inserts. It is found that a substantial reduction of surface roughness of 15.86% and 23.21% has been noticed. It may be attributed to the reduction of chip–tool interface adhesion, which substantially reduces the tool wear. Ze et al. [138] investigated the machinability of Ti alloy using textured cutting tools. A substantial reduction of coefficient of friction with rise of cutting speed was observed under dry cutting conditions. Greater reduction in the coefficient of friction has been evident for the rake surface textured cutting tool as compared to the flank surface textured tool and the conventional tool. Deng et al. [139] fabricated nano-textured cutting tools using a femtosecond laser with WS_2 solid lubricant coating. Machining tests have been carried out using rake face textured tools (TT), rake face textured tools deposited with WS_2 coatings (TT-WS_2) and conventional carbide tools (CT) under dry environment. There is a significant reduction of the cutting forces, the cutting temperature and the friction coefficient at the tool–chip interface of the TT and TT-WS_2 tools as compared to the conventional CT. The SEM micrographs of the femtosecond laser-textured rake face deposited with WS_2 solid lubricant coatings are shown in Figure 1.51 [139]. The WS_2 film fills both the space of ripples and the places around the ripples. The TT-WS_2 tool both with textured rake face and WS_2 coatings on its rake face shows the smallest cutting force among all the tools tested

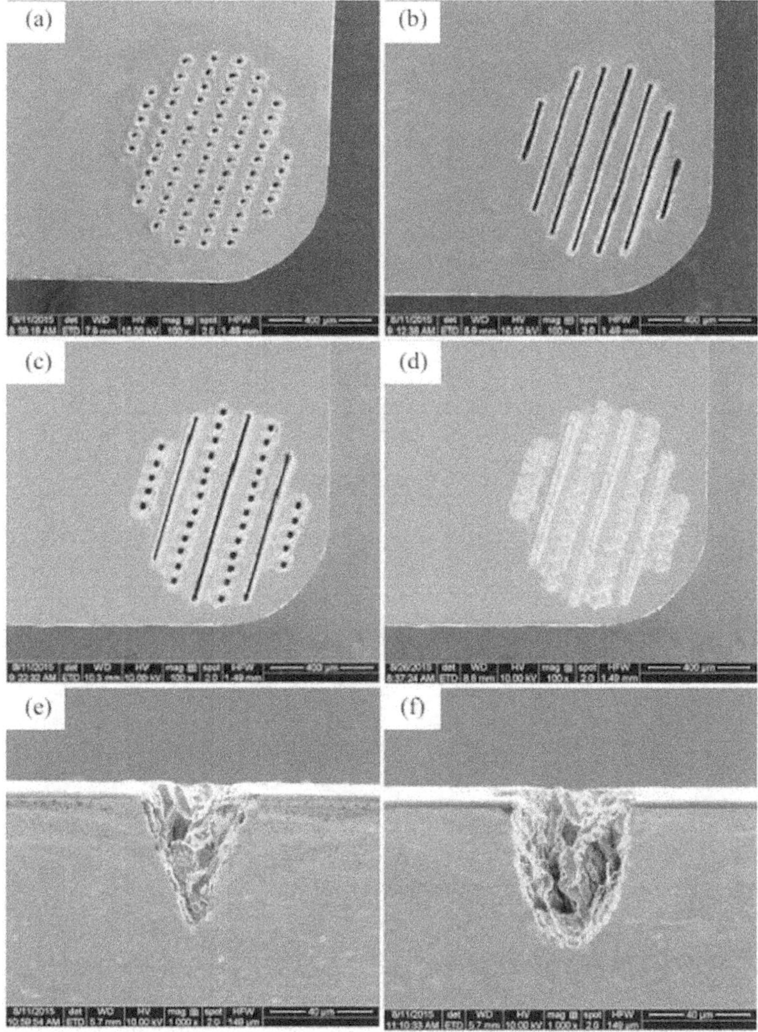

FIGURE 1.47 Micrographs of surface textures on the rake face of a carbide insert: (a) micro-groove texture type (GT), (b) micro-pit texture type (PT) and (c) hybrid textures (GPT); (d) micrograph of a GPT tool rake face filled with lubricant; cross-section micrograph of (e) a micro-groove and (f) a micro-pit. (From Ref. [132].)

under the same test conditions, as shown in Figure 1.52 [139]. The average coefficient of friction has been reduced by 13–26% using the TT-WS_2 as compared to the CT and TT, as is evident from Figure 1.53 [139].

Wenlong et al. [140] conducted machining experiments to investigate the performance of textured cutting tools. It is evident from the experiments that a graphite solid lubricant assisted cutting tool induces a lower coefficient of friction than MoS_2

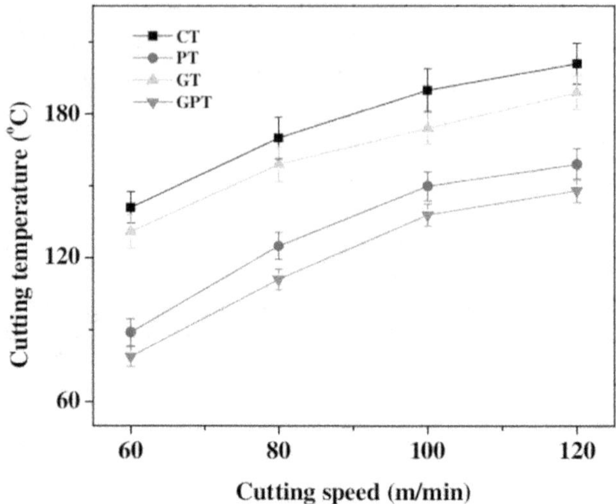

FIGURE 1.48 Cutting temperature of four kinds of tools at different speeds. (From Ref. [132].)

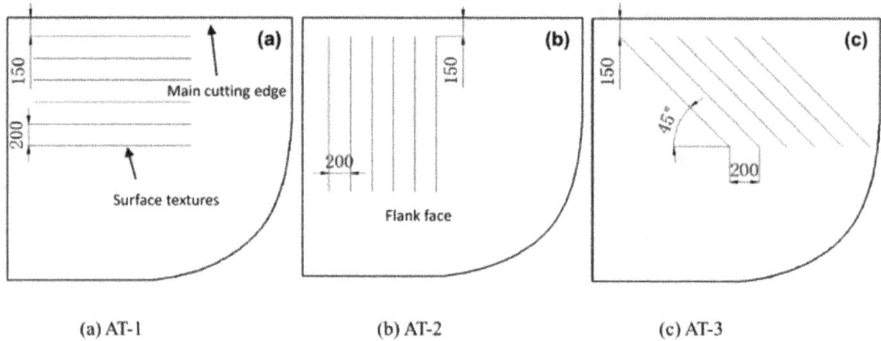

FIGURE 1.49 Diagrams of the flank-face textured tools (unit: μm): (a) textures parallel to the main cutting edge (AT-1), (b) textures perpendicular to the main cutting edge (AT-2) and (c) textures with an inclination angle of 45° to the cutting edge (AT-3). (From Ref. [135].)

and CaF_2 solid lubricant due to the formation of a thin lubricating film. Feng et al. [141] concluded that a reduction of cutting temperature of 10.1–12.3% occurred in a longitudinal textured cutting tool as compared to a conventional cutting tool during machining. As a result of this, the life of the cutting tool increases. Orra et al. [142] conducted hard machining using various textured cutting inserts such as horizontal micro-texture, vertical micro-texture and elliptical micro-texture with MoS_2 solid lubricants at the rake surface, as shown in Figure 1.54. It is observed from the experimental results that the coefficient of friction reduces for textured cutting tools as

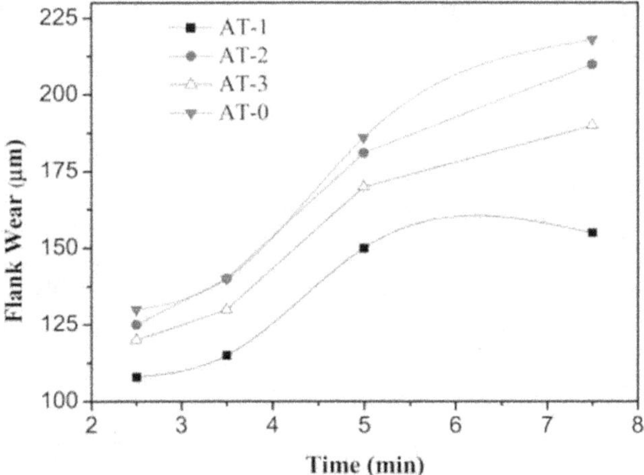

FIGURE 1.50 Flank wear of the tools with different textures in dry cutting of green alumina ceramics (v = 120 m/min, ap = 0.8 mm and f = 0.051 mm/r). (From Ref. [135].)

compared to non-textured inserts. As a result, cutting force decreases and shear angle of the cutting tool increases.

Farooqi and Yusoff [143] fabricated a novel textured cutting tool through EDM and investigated its performance under dry machining, which is of interest in green sustainable manufacturing. The investigation was based on the evolution of tool wear, surface roughness and cutting force and compared the performance with that of a non-textured tool. Fan et al. [144] fabricated five types of texture on the rake face of a PCBN cutting insert through a fiber laser technique, namely elliptical grooves, composite grooves, wavy grooves, transverse grooves and circular pits. Chip–tool contact area and cutting force increase with increased cutting speed, whereas shear angle decreases with increased cutting speed. The machining performance of the elliptical groove and wavy groove textured tools have been observed to be better than other textured tools under similar conditions, such as smaller contact area of chip and tool, cutting force, surface roughness and larger shear angle etc. There is a reduction of surface roughness by 11.73% to 56.7% because of the presence of a micro-textured cutting tool. Li et al. [145] investigated hard machining of GCr15 using micro-hole textured PCBN cutting inserts, and they performed well as compared to the non-textured cutting tool with respect to responses such as tool wear, surface roughness and cutting force. Sawant et al. [146] fabricated different textured cutting tools through the μ-PTAPD (i.e. μ-plasma power, exposure time and flow rate of powder) experimental apparatus used for the texturing on high-speed steel (HSS) cutting tools, as shown in Figure 1.55, and studied the performance of spot-textured, dimple-textured and non-textured HSS tools during machining of Ti-6Al-4V alloy with respect to surface roughness, flank wear, tool temperature, machining force and chip shape. The machinability performance of the spot-textured cutting tool outperformed the other textured cutting tools in reduction of cutting forces and coefficient of friction etc., as

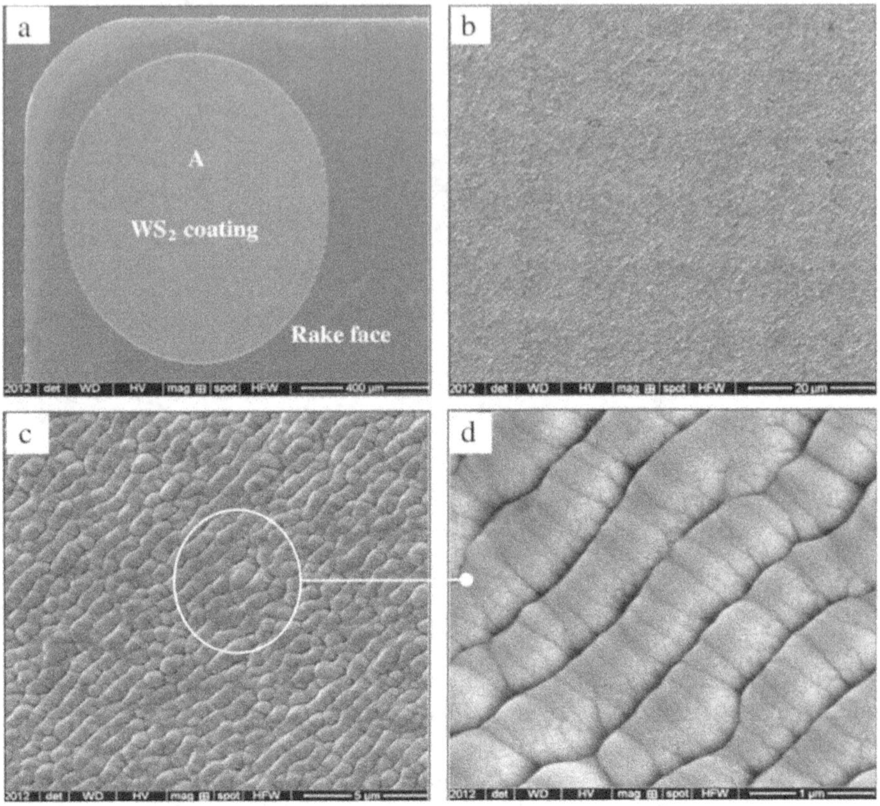

FIGURE 1.51 SEM micrographs of the femtosecond laser-textured rake face deposited with WS_2 solid lubricant coatings: (a) WS_2 coating on the femtosecond laser-textured rake face, (b) enlarged view marked in point A of (a), (c) enlarged view corresponding to (b), and (d) enlarged view corresponding to (c) showing fine details of the periodic pattern. (From Ref. [139].)

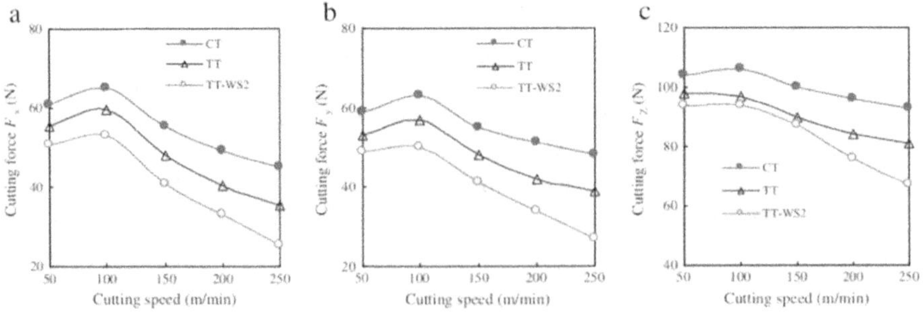

FIGURE 1.52 Cutting forces of CT, TT and TT-WS_2 tools at different cutting speeds: (a) axial thrust force Fx, (b) radial thrust force Fy, and (c) main force Fz (ap = 0.3 mm, f = 0.1 mm/r). (From Ref. [139].)

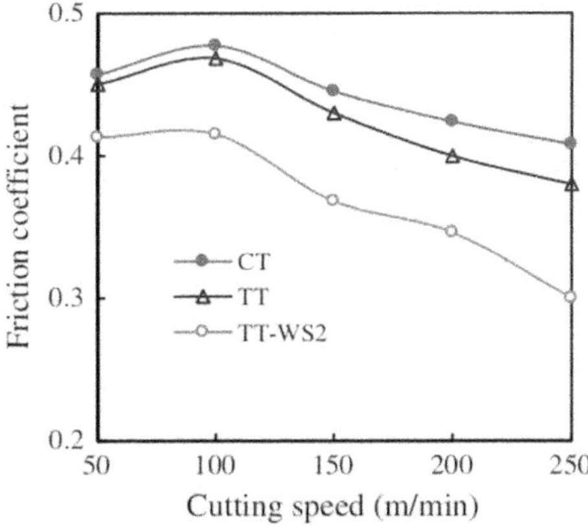

FIGURE 1.53 Friction coefficient between the tool–chip interface of CT, TT and TT-WS_2 tools at different cutting speeds (ap = 0.5 mm, f = 0.1 mm/r). (From Ref. [139].)

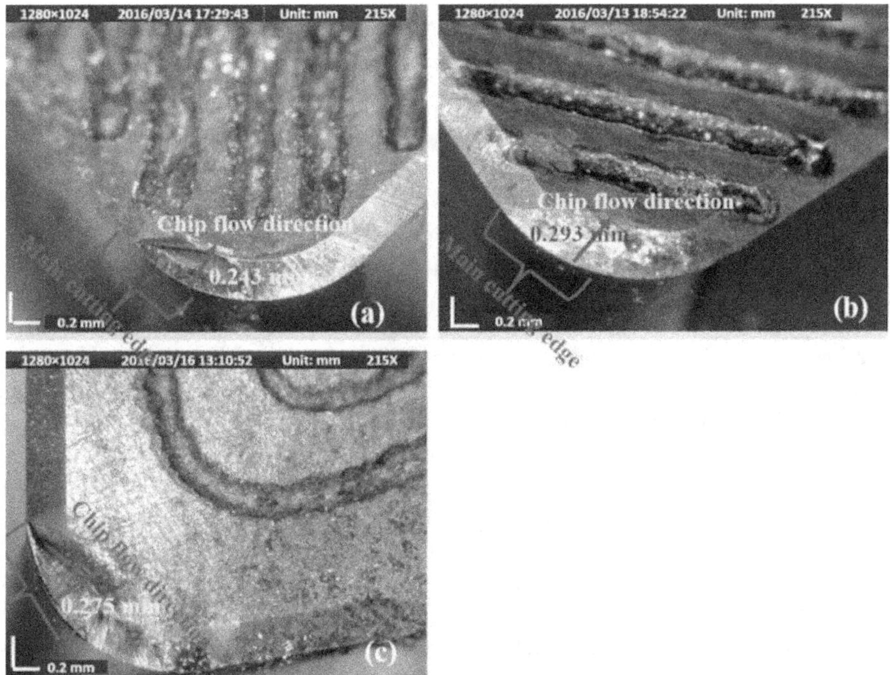

FIGURE 1.54 Image of micro-texture on the cutting insert indicating main cutting edge and chip flow direction after machining (a) vertical texture WV, (b) horizontal texture WH and (c) elliptical texture WE. (From Ref. [142].)

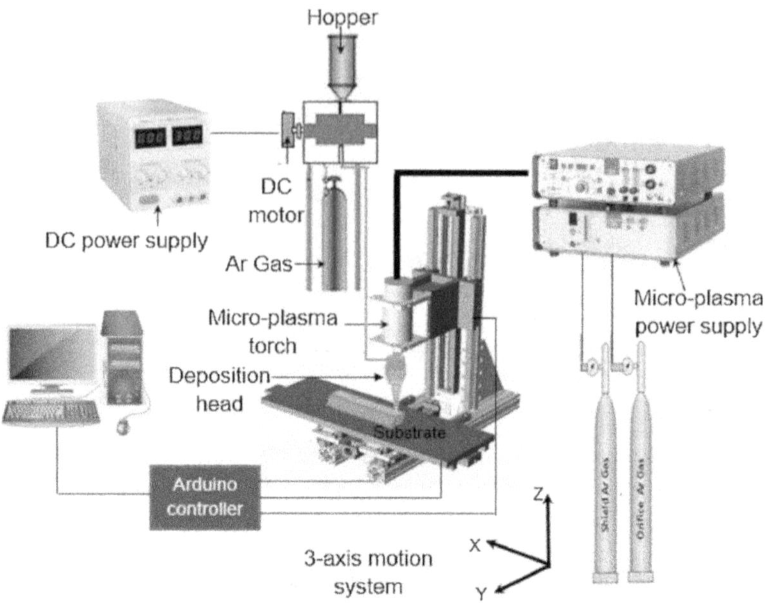

FIGURE 1.55 Schematic of the experimental apparatus of the μ-PTAPD process used in texturing of HSS cutting tools. (From Ref. [146].)

shown in Figure 1.56 and Figure 1.57 respectively [146] and proved to be effective, economical and environmentally friendly.

Ma et al. [147] investigated machining performance of mild steel using a micro-groove textured cutting tool and observed it to be better at lowering cutting force and energy requirements. Song et al. [148] fabricated a micro-hole texture on a carbide insert through micro-EDM with graphite embedded into the micro-holes so as to be self-lubricating. The micrographs of the micro-holes on the carbide rake face embedded without and with graphite are exhibited in Figure 1.58. Carbide inserts embedded with graphite (ST) and a conventional insert without micro-holes and graphite were utilized for comparison purposes during hard machining. The performance of the micro-hole textured cutting tool was found to be better at reducing tool wear and cutting temperature as compared to the non-textured cutting tool during the dry machining of hardened AISI 1045 steel, as is evident from Figure 1.59 and Figure 1.60 respectively [148].

Fang and Obikawa [149] conducted machining of Inconel 718 under high pressure jet cooling through a micro-textured cutting tool and observed it to be better as compared to a non-textured cutting tool in terms of reducing the tool wear. Su et al. [150] observed better machinability performance of micro-textured polycrystalline diamond cutting tools as compared to a non-textured cutting tool with respect to cutting force, coefficient of friction and anti-adhesion characteristics under dry conditions. Fibre laser techniques were employed to fabricate the micro-grooves. Several researchers in the recent past carried out machinability studies using textured

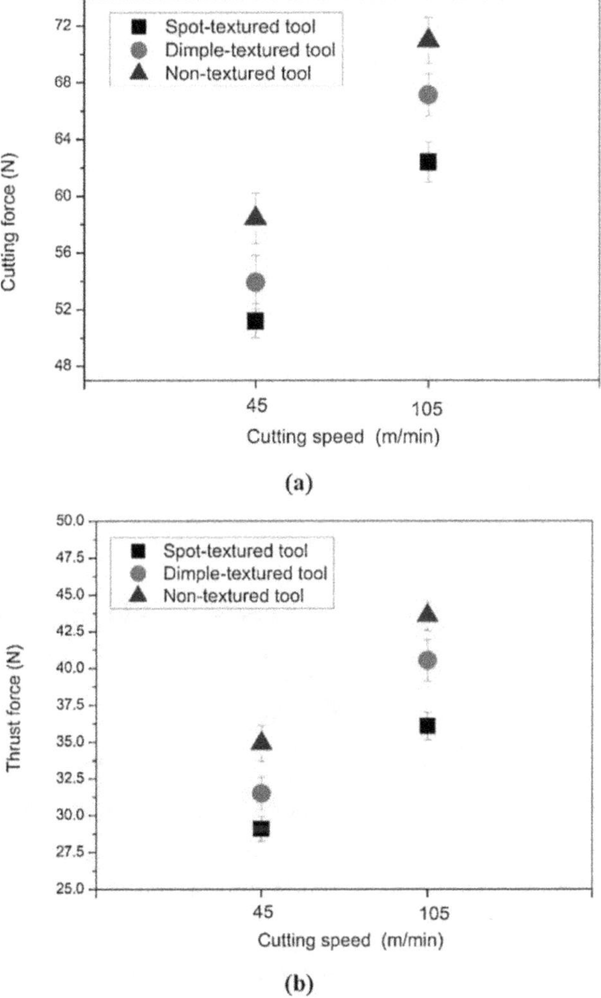

FIGURE 1.56 (a) Cutting force and (b) thrust force at different cutting speeds during turning of Ti-6Al-4V using spot-textured, dimple-textured and non-textured HSS tools. (From Ref. [146].)

cutting tools for difficult-to-cut materials and found them to be better because of enhanced tribological properties [151–158].

1.5 INFERENCES

Dry hard machining is the preferred method from an environmental point of view as it eliminates the use of conventional fluids and is safe for workers. However, it generates a large amount of heat, which is advantageous for machining hard materials as it

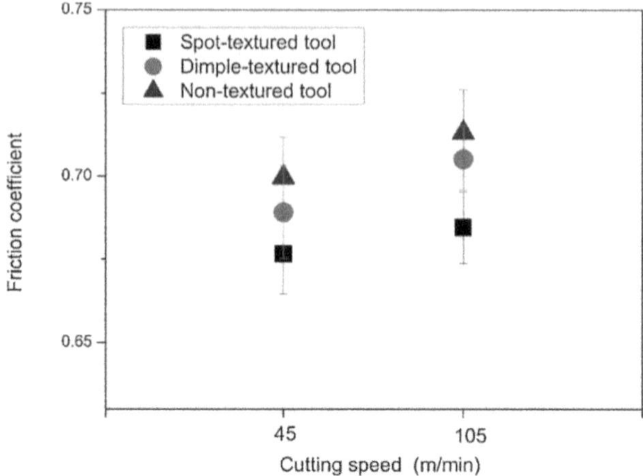

FIGURE 1.57 Coefficient of friction at different cutting speeds during turning of Ti-6Al-4V using spot-textured, dimple-textured and non-textured HSS tools. (From Ref. [146].)

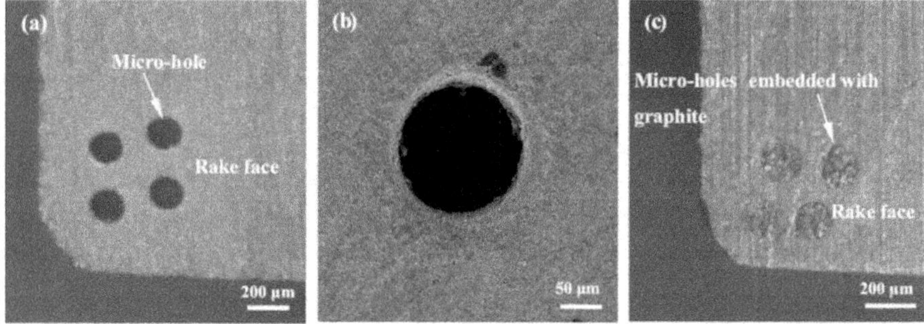

FIGURE 1.58 Micrographs of the micro-holes on the carbide rake face: (a) micro-holes filled without graphite, (b) enlarged micrograph corresponding to (a), and (c) micro-holes filled with graphite. (From Ref. [148].)

softens them and makes them easier to shear, but at the same time, it is detrimental because of the rapid growth of tool wear and tool failure. Hence further investigation is needed to eliminate the limitations so that it will be suitable for manufacturing industries. From an extensive review of the literature, it has been observed that coated carbide inserts have the potential in hard machining for tool life enhancement through the MQL technique. Therefore, more research is needed for eco-friendly alternatives to conventional cutting fluids in hard machining.

Sustainability of different cooling and lubrication strategies such as dry environment, MQL, NFMQL, cryogenics, high pressure cooling, hybrid cooling, use of textured cutting tools etc. in machining have been reviewed to eliminate the use of conventional cutting fluids and have been observed to be effective for environmental

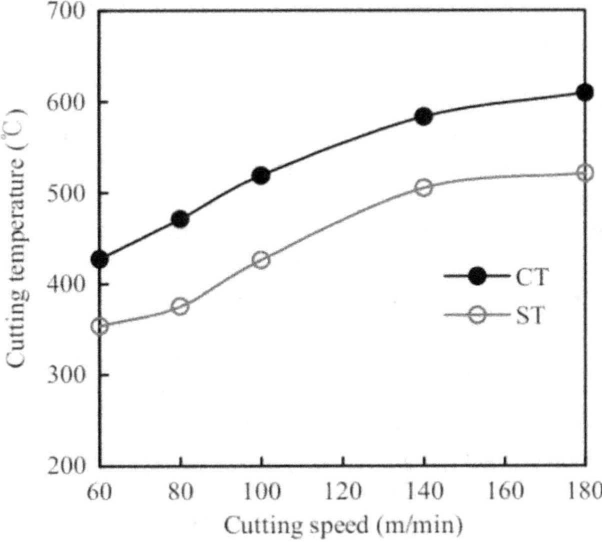

FIGURE 1.59 Cutting temperature of CT and ST tools in dry cutting of hardened steel at different cutting speeds (ap = 0.2 mm, f = 0.1 mm/r, cutting time 5 min). (From Ref. [148].)

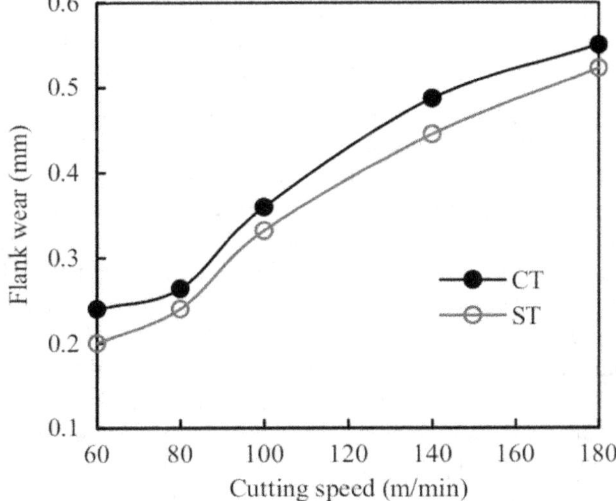

FIGURE 1.60 Flank wear of CT and ST tools in dry cutting of hardened steel (ap = 0.2 mm, f = 0.1 mm/r, cutting time 5 min). (From Ref. [148].)

protection and energy consumption savings. Substantial reductions in manufacturing costs with enhancement of tool life have also been noticed. In general, MQL, NFMQL and hybrid cooling improve overall sustainability in terms of economic, technical, environmental and operator health aspects.

It has been noticed from the literature that limited studies are undertaken using coolants in hard machining and most prefer to operate under dry conditions. There is agreement and disagreement among researchers about the application of coolants in machining hardened steel, and it is still open for future studies. Limited research has been carried out for sustainable machining of hardened steel. This review will be immensely beneficial for researchers on sustainable hard machining and shop floors as well. Hard machining can be made economically, ecologically, societally and technologically effective by utilizing appropriate machining parameters, geometrical parameters, cutting tool materials, tool coatings and environmental parameters to achieve sustainable machining or clean/green machining through appropriate sustainability assessment and cooling/lubrication techniques. Investigation is still necessary for the synthesis and application of hybrid nanofluids in hard machining for the enhancement of machinability and sustainability. The investigation should be focused on types of nanofluids, nanoparticle size and shape, their concentration, the life and biodegradability of nanofluids, flow rate, nozzle orientation angle, spray standoff distance and air pressure, as well as their effects on hard machining responses, in order to improve efficiency and create positive ecological enhancements. The optimization of these parameters for nanofluids is required for effective application in hard machining. Further, a life cycle assessment model for sustainability aspects is worthwhile for investigation in terms of environmental, economic and social impacts etc. Additional features such as human health, atmospheric pollution and vibration effects may be studied in future work. Investigation into the environmental and toxicity effects of nanoparticles in hard machining applications through MQL is needed. The hybrid nanoparticles are very beneficial for improving the lubricity of MQL lubricants, and this needs detailed investigation in hard machining. Sustainability assessment along with thermal analysis of MQL lubricants may be assessed for clean/green manufacturing. More studies are to be undertaken for the development of tribo-film formation models, analysis of heat transfer by MQL, nanoparticle suspensions, wetting ability of droplets etc. In the literature, nanofluid cooling and lubrication has been used through the MQL technique. There is scope to implement it through the spray impingement cooling method for comparison with the MQL technique. More research is needed with integration of computational algorithms and data science in MQL enabled hard machining in order to come up with intelligent and smart manufacturing for Industry 4.0. It is observed that the future direction of sustainable machining is in the application of nanoparticles and their formation.

REFERENCES

[1] G. Bartarya, S.K. Choudhury, State of the art in hard turning. *International Journal of Machine Tools & Manufacture*, 53, 1–14, 2012.

[2] D. Huddle, New hard turning tools and techniques offer a cost-effective alternative to grinding. *Tooling and Production Magazine*, 2001.

[3] W. Grzesik, Machining of hard materials. In Davim, J. P. (ed.), *Machining*, p. 100. Springer, London, 2008. https://link.springer.com/chapter/10.1007/978-1-84800-213-5_4

[4] H. Tönshoff, C. Arendt, R. Ben Amor, Cutting of hardened steel, *Annals of the CIRP, Keynote Paper*, 49(2), 547–565, 2000.

[5] S. K. Choudhury, S. Chinchanikar, Finish machining of hardened steel, comprehensive materials finishing, *Reference Module in Materials Science and Materials Engineering*, 1, 47–92, 2017.

[6] R. Suresh, S. Basavarajappa, V. N Gaitonde, G. L. Samuel, J. P. Davim, State-of-the-art research in machinability of hardened steels. Proceedings of the Institution of Mechanical Engineers, Part B, 227(2), 191–209, 2013.

[7] V. P. Astakhov, *Machining of Hard Materials – Definitions and Industrial Applications, Chapter, Machining of Hard Materials*. Springer-Verlag London Ltd., 1–32, 2011.

[8] Modern Machine Shop. www.mmsonline.com/articles/hard-turning-might-not-be-as-hardas-you-think

[9] W. Grezesik, T. Wanat, Surface finish generated in HT of quenched alloy steel parts using conventional and wiper ceramic inserts. *International Journal of Machine Tools and Manufacture,* 46, 1988–1995, 2006.

[10] V. Dessoly, S. N. Melkote, C. Lescalier, Modeling and verification of cutting tool temperature in rotary tool turning of hardened steels. *International Journal of Machine Tools and Manufacture,* 44, 1463–1470, 2004.

[11] A. S. More, W. Jiang, W. D. Brown, A. P. Malshe, Tool wear and machining performance of cBN–TiN coated carbide inserts and PCBN compact inserts in turning AISI 4340 hardened steel. *Journal of Materials Processing Technology,* 180, 253–262, 2006.

[12] Z. Zurecki, R. Ghosh, J. H. Frey, Investigation of white layer formed in conventional and cryogenic hard turning of steel. In: Proceedings of ICEME'03, ASME International. Mechanical Engineering Congress and Exposition, Washington D.C., November 16–21, 2003.

[13] A. Noorul Haq, T. Tamizharasan, Investigation of the effects of cooling in hard turning operations. *International Journal of Advanced Manufacturing Technology,* 30, 808–816, 2006.

[14] A. K. Sahoo, B. Sahoo, Experimental investigations on machinability aspects in finish hard turning of AISI 4340 steel using uncoated and multilayer coated carbide inserts. *Measurement,* 45, 2153–2165, 2012.

[15] R. Suresh, S. Basavarajappa, V. N. Gaitonde, G. L. Samuel, Machinability investigations on hardened AISI 4340 steel using coated carbide insert. *International Journal of Refractory Metals and Hard Materials,* 33, 75–86, 2012.

[16] B. Sen, M. K. Gupta, M. Mia, U. K. Mandal, S. P. Mondal, Wear behaviour of TiAlN coated solid carbide end-mill under alumina enriched minimum quantity palm oil-based lubricating condition. *Tribology International*, 148, 106310, 2020.

[17] M. K. Gupta, Q. Song, Z. Liu, M. Sarikaya, M. Jamil, M. Mia, A. K. Singla, A. M. Khan, N. Khanna, D. Y. Pimenov, Environment and economic burden of sustainable cooling/lubrication methods in machining of Inconel-800. *Journal of Cleaner Production*, 287, 125074, 2020. https://doi.org/10.1016/j.jclepro.2020.125074.

[18] H. Salvi, H. Vesuwala, P. Raval, V. Badheka, N. Khanna, Sustainability analysis of additive + subtractive manufacturing processes for Inconel 625. *Sustainable Materials and Technologies*, 35, e00580, 2023.

[19] L. S. R. Krishna, P. J. Srikanth, Evaluation of environmental impact of additive and subtractive manufacturing processes for sustainable manufacturing. *Materials Today: Proceedings*, 3054–3060, 2021. https://doi.org/10.1016/j. matpr.2020.12.060.

[20] C. G. Machado, M. P. Winroth, E. H. D. Ribeiro da Silva, Sustainable manufacturing in industry 4.0: an emerging research agenda. *International Journal of Production Research*, 58, 1462–1484, 2020. https://doi.org/10.1080/00207543.2019.1652777.

[21] G. Kshitij, N. Khanna, Ç.V. Yıldırım, S. Daglı, M. Sarıkaya, Resource conservation and sustainable development in the metal cutting industry within the framework of the green economy concept: An overview and case study. *Sustainable Materials Technology*, 34, e00507, 2022. https://doi.org/10.1016/j.susmat.2022.e00 507 e00507.

[22] V. T. Le, H. Paris, G. Mandil, Environmental impact assessment of an innovative strategy based on an additive and subtractive manufacturing combination. *Journal of Cleaner Production*, 164, 508–523, 2017. https://doi.org/10.1016/j. jclepro.2017.06.204.

[23] S. K. Tamang, M. Chandrasekaran, A. K. Sahoo, Sustainable machining: An experimental investigation and optimization of machining Inconel 825 with dry and MQL approach. *Journal of the Brazilian Society of Mechanical Sciences and Engineering*, 40, 374, 2018.

[24] A. Zein, W. Li, C. Herrmann, S. Kara, Energy efficiency measures for the design and operation of machine tools: An axiomatic approach. In: 18th CIRP International Conference on Life Cycle Engineering: Globalized Solutions for Sustainability in Manufacturing, *Braunschweig, Germany*, 274–279, 2011. https://doi.org/10.1007/ 978-3-642-19692-8_48.

[25] K. M. Rajan, A. K. Sahoo, B. C. Routara, R. Kumar, Investigation on surface roughness, tool wear and cutting power in MQL turning of bio-medical Ti-6Al-4V ELI alloy with sustainability. Proceedings of the Institution of Mechanical Engineers, Part E, 236(4), 1452–1466, 2022.

[26] A. Shokrani, V. Dhokia, S. T. Newman, Environmentally conscious machining of difficult-to-machine materials with regard to cutting fluids. *International Journal of Machine Tools & Manufacture*, 57, 83–101, 2012.

[27] G. M. Krolczyk, R. W. Maruda, J. B. Krolczyk, S. Wojciechowski, M. Mia, P. Nieslony, G. Budzik, Ecological trends in machining as a key factor in sustainable production: A review. *Journal of Cleaner Production,* 218, 601e615, 2019.

[28] W. Grzesik, Nanofluid assistance in machining processes-properties, mechanisms and applications: A review. *Journal of Machine Engineering*, 21(2), 75–90, 2021.

[29] A. Panda, A. K. Sahoo, A. K. Rout, Investigations on surface quality characteristics with multi-response parametric optimization and correlations. *Alexandria Engineering Journal* 55, 1625–1633, 2016.

[30] M. Lahres, O. Doerfel, R. Neumüller, Applicability of different hard coatings in dry machining an austenitic steel. *Surface and Coatings Technology*, 120–121, 687–691, 1999.

[31] A. Panda, A. K. Sahoo, I. Panigrahi, A. K. Rout, Prediction models for on-line cutting tool and machined surface condition monitoring during hard turning considering vibration signal. *Mechanics & Industry*, 21, 520, 2020.

[32] A. E. Diniz, R. Micaroni, Cutting conditions for finish turning process aiming: The use of dry cutting. *International Journal of Machine Tools and Manufacture,* 42(8), 899–904, 2002.
[33] A. I. Fernandez-Abia, J. Barreiro, L. N. L. d. Lacalle, S. Martínez, Effect of very high cutting speeds on shearing, cutting forces and roughness in dry turning of austenitic stainless steels. *International Journal of Advanced Manufacturing Technology,* 57(1), 61–71, 2011.
[34] R. Kumar, A. K. Sahoo, R. K. Das, A. Panda, P. C. Mishra, Modelling of flank wear, surface roughness and cutting temperature in sustainable hard turning of AISI D2 steel, *Procedia Manufacturing*, 20, 406–413, 2018.
[35] A. K. Sahoo, B. Sahoo, Performance studies of multilayer hard surface coatings (TiN/TiCN/Al2O3/TiN) of indexable carbide inserts in hard machining: Part-II (RSM, grey relational and techno economical approach). *Measurement*, 46(8), 2868–2884, 2013.
[36] A. K. Sahoo, B. Sahoo, A comparative study on performance of multilayer coated and uncoated carbide inserts when turning AISI D2 steel under dry environment. *Measurement*, 46(8), 2695–2704, 2013.
[37] Y.-W. Park, Tool material dependence of hard turning on the surface quality. *International Journal of Korean Society of Precision Engineering,* 3(1), 76–82, 2002.
[38] F. Klocke, G. Eisenblatter, Dry cutting. *CIRP Annals—Manufacturing Technology,* 46, 519–526, 1997.
[39] S. Chinchanikar, S. K. Choudhury, Machining of hardened steel—Experimental investigations, performance modeling and cooling techniques: A review. *International Journal of Machine Tools and Manufacture,* 89, 95–109, 2015.
[40] S. K. Tamang, M. Chandrasekaran, A. K. Sahoo, Sustainable machining: an experimental investigation and optimization of machining Inconel 825 with dry and MQL approach. *Journal of the Brazilian Society of Mechanical Sciences and Engineering*, 40, 374, 2018.
[41] M. Mia, M. K. Gupta, G. Singh, G. Królczyk, D. Y. Pimenov, An approach to cleaner production for machining hardened steel using different cooling-lubrication conditions. *Journal of Cleaner Production*, 187, 1069–1081, 2018.
[42] L. Dash, S. Padhan, S. R. Das, Experimental investigations on surface integrity and chip morphology in hard tuning of AISI D3 steel under sustainable nanofuid-based minimum quantity lubrication. Journal of the *Brazilian Society of Mechanical Sciences and Engineering*, 42, 500–525, 2020.
[43] S. Padhan, A. Das, A. Santoshwar, T. R. Dharmendrabhai, S. R. Das, Sustainability assessment and machinability investigation of austenitic stainless steel in finish turning with advanced ultra-hard SiAlON ceramic tool under different cutting environments. *Silicon*, 13, 119–147, 2021.
[44] A. Panda, S. R. Das, D. Dhupal, Machinability investigation and sustainability assessment in FDHT with coated ceramic tool. *Steel and Composite Structures*, 34(5), 681–698, 2020.
[45] A.T. Abbas, F. Benyahia, M. M. El Rayes, C. Pruncu, M. A. Taha, H. Hegab, Towards optimization of machining performance and sustainability aspects when turning AISI 1045 steel under different cooling and lubrication strategies, *Materials*, 12, 3023–3049, 2019.
[46] A. T Abbas, M. K. Gupta, M. S. Soliman, M. Mia, H. Hegab, M. Luqman, D. Y. Pimenov, Sustainability assessment associated with surface roughness and power consumption characteristics in nanofluid MQL-assisted turning of AISI 1045

steel, *The International Journal of Advanced Manufacturing Technology*, 105, 1311–1327, 2019.

[47] K. K. Gajrani, P. S. Suvin, S. V. Kailas, M. R. Sankar, Thermal, rheological, wettability and hard machining performance of MoS_2 and CaF_2 based minimum quantity hybrid nano-green cutting fluids, *Journal of Materials Processing Technology*, 266, 125–139, 2019.

[48] P. S. Sreejith, Machining of 6061 aluminium alloy with MQL, dry and flooded lubricant conditions. *Materials Letters*, 62(2), 276–278, 2008.

[49] S. Masoudi, M. J. Esfahani, F. Jafarian, S. A. Mirsoleimani, Comparison the effect of MQL, Wet and Dry turning on surface topography, cylindricity tolerance and sustainability, *International Journal of Precision Engineering and Manufacturing-Green Technology*, 2019. https://doi.org/10.1007/s40684-019-00042-3

[50] M. Goldberg, Improving productivity by using innovative metal cutting solutions with an emphasis on green machining. *International Journal of Machining and Machinability of Materials*, 12 (1/2), 117–125, 2012.

[51] V. Upadhyay, P. K. Jain, N. K. Mehta, Machining with minimum quantity lubrication: a step towards green manufacturing. *International Journal of Machining and Machinability of Materials*, 13(4), 349–371, 2013.

[52] M. G. Faga, P. C. Priarone, M. Robiglio, L. Settineri, V. Tebaldo, Technological and sustainability implications of dry, near-dry, and wet turning of Ti-6Al-4V alloy. *International Journal of Precision Engineering and Manufacturing-Green Technology*, 4, 129–139, 2017.

[53] Y. Shokoohi, E. Khosrojerdi, B. h. R. Shiadhi, Machining and ecological effects of a new developed cutting fluid in combination with different cooling techniques on turning operation. *Journal of Cleaner Production*, 94, 330–339, 2015.

[54] S. Debnath, M. M. Reddy, Q. S. Yi, Environmental friendly cutting fluids and cooling techniques in machining: A review. *Journal of Cleaner Production*, 83, 33–47, 2014.

[55] I. Deiab, S. W. Raza, S. Pervaiz, Analysis of lubrication strategies for sustainable machining during turning of titanium Ti-6Al-4V alloy. *Procedia CIRP*, 17, 766–771, 2014.

[56] P. C. Priarone, M. Robiglio, L. Settineri, V. Tebaldo, Effectiveness of minimizing cutting fluid use when turning difficult-to-cut alloys. *Procedia CIRP*, 29, 341–346, 2015.

[57] M. K. Gupta, M. Mia, G. R. Singh, D. Y. Pimenov, M. Sarikaya, V. S. Sharma, Hybrid cooling-lubrication strategies to improve surface topography and tool wear in sustainable turning of Al 7075-T6 alloy. *The International Journal of Advanced Manufacturing Technology*, 101, 55–69, 2019.

[58] O. Pereira, J. E. M. Alfonso, A. Rodríguez, A. Calleja, A. F. Valdivielso, L. N. L. de Lacalle, Sustainability analysis of lubricant oils for minimum quantity lubrication based on their tribo-rheological performance. *Journal of Cleaner Production*, 164, 1419–1429, 2017.

[59] R. K. Das, A. K. Sahoo, P. C. Mishra, R. Kumar, A. Panda, Comparative machinability performance of heat treated 4340 Steel under dry and minimum quantity lubrication surroundings. *Procedia Manufacturing*, 20, 377–385, 2018.

[60] R. Kumar, A. K. Sahoo, P. C. Mishra, R. K. Das, Influence of Al2O3 and TiO2 nanofluid on hard turning performance. *The International Journal of Advanced Manufacturing Technology*, 106, 2265–2280, 2020.

[61] S. Roy, R. Kumar, A. K. Sahoo, R. K. Das, A brief review on effects of conventional and nanoparticle based machining fluid on machining performance of minimum quantity lubrication machining, *Materials Today: Proceedings*, 18(7), 5421–5431, 2019.

[62] R. K. Das, R. Kumar, G. Sarkar, S. Sahoo, A. K. Sahoo, P. C. Mishra, Comparative machining performance of hardened AISI 4340 steel under dry and minimum quantity lubrication environments. *Materials Today: Proceedings*, 5(11), 24898–24906, 2018.

[63] S. Dambhare, S. Deshmukh, A. Borade, A. Digalwar, M. Phate, Sustainability issues in turning process: A study in indian machining industry, *Procedia CIRP*, 26, 379–384, 2015.

[64] A. Das, S. K. Patel, B. B. Biswal, S. R. Das, Performance evaluation of aluminium oxide nano particles in cutting uid with minimum quantity lubrication technique in turning of hardened AISI 4340 alloy steel. *Scientia Iranica B*, 27(6), 2838–2852, 2020.

[65] M. A. ul. Haq, S. Hussain, M. A. Ali, M. U. Farooq, N. A. Mufti, C. I. Pruncud, A. Wasim, Evaluating the effects of nano-fluids based MQL milling of IN718 associated to sustainable productions, *Journal of Cleaner Production*, 310, 127463, 2021.

[66] J. P. Davim, P. S. Sreejith, J. Silva, Some studies about machining of MMC's by MQL (Minimum Quantity of Lubricant) conditions. *Advanced Composites Letters*, 18(1), 21–23, 2009.

[67] A. Tayal, N. S. Kalsi, M. K. Gupta, A. G. Collado, M. Sarikaya, Reliability and economic analysis in sustainable machining of Monel 400 alloy. Proceedings of the Institution of Mechanical Engineers, Part C: *Journal of Mechanical Engineering Science*, 2021. https://doi.org/10.1177/0954406220986818.

[68] M. K. Gupta, Q. Song, Z. Liu, M. Sarikaya, M. Jamil, M. Mia, A. K. Singla, A. M. Khan, N. Khanna, D. Y. Pimenov, Environment and economic burden of sustainable cooling/lubrication methods in machining of Inconel-800, *Journal of Cleaner Production*, 287, 125074, 2021.

[69] R. K. Das, A. K. Sahoo, R. Kumar, S. Roy, P. C. Mishra, T. Mohanty, MQL assisted cleaner machining using PVD TiAlN coated carbide insert: Comparative assessment. *Indian Journal of Engineering & Materials Sciences*, 26, 311–325, 2019.

[70] K. Venugopal, S. Paul, A. Chattopadhyay, Tool wear in cryogenic turning of Ti–6Al–4V alloy. *Cryogenics,* 47, 12–18, 2007.

[71] E. Abele, B. Schramm, Using PCD for machining CGI with a CO2 coolant system, *Production Engineering,* 2, 165–169, 2008.

[72] F. Pusavec, D. Kramar, P. Krajnik, J. Kopac, Transitioning to sustainable production—part II: Evaluation of sustainable machining technologies, *Journal of Cleaner Production,* 18, 1211–1221, 2010.

[73] M. J. Bermingham, J. Kirsch, S. Sun, S. Palanisamy, M. S. Dargusch, New observations on tool life, cutting forces and chip morphology in cryogenic machining Ti–6Al–4V. *International Journal of Machine Tools and Manufacture,* 51, 500–511, 2011.

[74] F. Pusavec, J. Kopac, Achieving and implementation of sustainability principles in machining processes. *Journal of Advances in Production Engineering and Management,* 4(3), 151–160, 2009.

[75] S. Hong, Lubrication mechanisms of LN2 in ecological cryogenic machining. *Machining Science and Technology,* 10, 33–155, 2006.

[76] M. Dhananchezian, M. P. Kumar, Cryogenic turning of the Ti–6Al–4V alloy with modified cutting tool inserts. *Cryogenics,* 51, 34–40, 2011.

[77] S. Y. Hong, Y. Ding, W. Jeong, Friction and cutting forces in cryogenic machining of Ti–6Al–4V. *International Journal of Machine Tools and Manufacture,* 41, 2271–2285, 2001.

[78] Y. Ding, J. Jeong, S. Hong, Experimental evaluation of friction coefficient and liquid nitrogen lubrication effect in cryogenic machining. *Machining Science and Technology,* 6, 235–250, 2002.

[79] C. Agrawal, J. Wadhwa, A. Pitroda, C. I. Pruncu, M. Sarikaya, N. Khanna, Comprehensive analysis of tool wear, tool life, surface roughness, costing and carbon emissions in turning Ti–6Al–4V titanium alloy: Cryogenic versus wet machining, *Tribology International*, 153, 106597, 2021.

[80] S. R. Nandam, U. Ravikiran, A. A. Rao, Machining of tungsten heavy alloy under cryogenic environment. *Procedia Materials Science,* 6, 296–303, 2014.

[81] M. Jamil, W. Zhao, N. He, M. K. Gupta, M. Sarikaya, A. M. Khan, M. R. Sanjay, S. Siengchin, D. Y. Pimenov, Sustainable milling of Ti–6Al–4V: A trade-off between energy efficiency, carbon emissions and machining characteristics under MQL and cryogenic environment. *Journal of Cleaner Production*, 281, 125374, 2021.

[82] S. Sun, M. Brandt, M. S. Dargusch, Machining Ti–6Al–4V alloy with cryogenic compressed air cooling. *International Journal of Machine Tools and Manufacture,* 50, 933–942, 2010.

[83] L. Brandao, R. Coelho, A. Rodrigues, Experimental and theoretical study of workpiece temperature when end milling hardened steels using (TiAl)Ncoated and PcBN-tipped tools, *Journal of Materials Processing Technology,* 199, 234–244, 2008.

[84] J. Gisip, R. Gazo, H. A. Stewart, Effects of cryogenic treatment and refrigerated air on tool wear when machining medium density fiberboard. *Journal of Materials Processing Technology,* 209, 5117–5122, 2009.

[85] S. Kim, D. Lee, M. Kang, J. Kim, Evaluation of machinability by cutting environments in high-speed milling of difficult-to-cut materials. *Journal of Materials Processing Technology,* 111 256–260, 2001.

[86] B. Yalcin, A. Ozgur, M. Koru, The effects of various cooling strategies on surface roughness and tool wear during soft materials milling. *Materials & Design,* 30, 896–899, 2009.

[87] J. Liu, Y. Kevinchou, On temperatures and tool wear in machining hypereutectic Al–Si alloys with vortex-tube cooling. *International Journal of Machine Tools and Manufacture,* 47, 635–645, 2007.

[88] Y. Su, N. He, L. Li, X. Li, An experimental investigation of effects of cooling/lubrication conditions on tool wear in high-speed end milling of Ti–6Al–4V. *Wear,* 261, 760–766, 2006.

[89] S. M. Yuan, L. T. Yan, W. D. Liu, Q. Liu, Effects of cooling air temperature on cryogenic machining of Ti–6Al–4V alloy. *Journal of Materials Processing Technology,* 211, 356–362, 2011.

[90] Y. Su, N. He, L. Li, A. Iqbal, M. Xiao, S. Xu, B. Qiu, Refrigerated cooling air cutting of difficult-to-cut materials. *International Journal of Machine Tools and Manufacture,* 47, 927–933, 2007.

[91] M. Rahman, A. S. Kumar, M. U. Salam, M. Ling, Effect of chilled air on machining performance in end milling. *The International Journal of Advanced Manufacturing Technology* 21, 787–795, 2003.

[92] J. Liu, Y. Kevinchou, On temperatures and tool wear in machining hypereutectic Al–Si alloys with vortex-tube cooling. *International Journal of Machine Tools and Manufacture,* 47, 635–645, 2007.

[93] F. A. Cotton, G. Wilkinson, C. A. Murillo, *Advanced Inorganic Chemistry*. 6th ed. New Delhi, India: John Wiley & Sons, 2004.

[94] P. V. Krishna, R. R. Srikant, D. N. Rao, Experimental investigation on the performance of nanoboric acid suspensions in SAE-40 and coconut oil during turning of AISI 1040 steel. *International Journal of Machine Tools and Manufacture,* 50, 911–916, 2010.

[95] N. S. K. Reddy, P. V. Rao, Experimental investigation to study the effect of solid lubricants on cutting forces and surface quality in end milling. *International Journal of Machine Tools and Manufacture,* 46(2), 189–198, 2006.

[96] S. Shaji, V. Radhakrishnan, An investigation on solid lubricant moulded grinding wheels. *International Journal of Machine Tools and Manufacture,* 43, 965–972, 2003.

[97] N. S. K. Reddy, P. V. Rao, Performance improvement of end milling using graphite as a solid lubricant. *Materials and Manufacturing Processes,* 20, 673–686, 2005.

[98] D. Singh, P. V. Rao, Performance improvement of hard turning with solid lubricants. *International Journal of Advanced Manufacturing Technology,* 38, 529–535, 2008.

[99] D. Mukhopadhyay, S. Banerjee, N. S. K. Reddy, Investigation to study the applicability of solid lubricant in turning AISI 1040 steel. *Journal of Manufacturing Science & Engineering*, 129, 520–526, 2007.

[100] S. Debnath, M. M. Reddy, Q. S. Yi, Environmental friendly cutting fluids and cooling techniques in machining: A review. *Journal of Cleaner Production,* 83, 33–47, 2014.

[101] M. Mia, P. R. Dey, M. S. Hossain, M. T. Arafat, M. Asaduzzaman, M. S. Ullah, S. M. T Zobaer, Taguchi S/N based optimization of machining parameters for surface roughness, tool wear and material removal rate in hard turning under MQL cutting condition. *Measurement,* 122, 380e391, 2018.

[102] E. O. Ezugwu, J. Bonney, R. B. Da Silva, O. Cakir, Surface integrity of finished turned Tie6Ale4V alloy with PCD tools using conventional and high pressure coolant supplies. *International Journal of Machine Tools and Manufacture,* 47(6), 884e891, 2007.

[103] R. Kumar, A. K. Sahoo, P. C. Mishra, R. K. Das, Measurement and machinability study under environmentally conscious spray impingement cooling assisted machining, *Measurement*, 135, 913–927, 2019.

[104] R. Kumar, A. K. Sahoo, P. C. Mishra, R. K. Das, S. Roy, ANN modeling of cutting performances in spray cooling assisted hard turning. *Materials Today: Proceedings*, 5(9), 18482–18488, 2018.

[105] P. C. Mishra, D. K. Das, M. Ukamanal, B. C. Routara, A. K. Sahoo, Multi-response optimization of process parameters using Taguchi method and grey relational analysis during turning AA 7075/SiC composite in dry and spray cooling environments. *International Journal of Industrial Engineering Computations*, 6(4), 445–456, 2015.

[106] S. K. Sahu, P. C. Mishra, K. Orra, A. K. Sahoo, Performance assessment in hard turning of AISI 1015 steel under spray impingement cooling and dry environment, IMechE, *Part B: Journal of Engineering Manufacture,* 229(2), 251–265, 2015.

[107] M. Ukamanal, P. C. Mishra, A. K. Sahoo, Temperature distribution during AISI 316 steel turning under TiO_2-water based nanofluid spray environments. *Materials Today: Proceedings*, 5(9), 20741–20749, 2018.

[108] M. Ukamanal, P. C. Mishra, A. K. Sahoo, Effects of spray cooling process parameters on machining performance AISI 316 Steel: A novel experimental technique, *Experimental Techniques*, 44, 19–36, 2020.

[109] R. Kumar, A. K. Sahoo, P. C. Mishra, R. K. Das, Influence of Al_2O_3 and TiO_2 nanofluid on hard turning performance. *International Journal of Advanced Manufacturing Technology,* 106, 2265–2280, 2020.

[110] B. Sen, M. Mia, G. M. Krolczyk, U. K. Mandal, S. P. Mondal, Eco-friendly cutting fluids in minimum quantity lubrication assisted machining: A review on the perception of sustainable manufacturing, eco-friendly cutting fluids in minimum quantity lubrication assisted machining: *A review on the perception of sustainable manufacturing. International Journal of Precision Engineering and Manufacturing-Green Technology*, 8, 249–290, 2019.

[111] Y. Su, N. He, L. Li, Effect of cryogenic minimum quantity lubrication (CMQL) on cutting temperature and tool wear in high-speed end milling of titanium alloys. In: *Applied Mechanics and Materials*, 34, 816–1821, 2010.

[112] S. Yuan, L. Yan, W. Liu, Q. Liu, Effects of cooling air temperature on cryogenic machining of Ti–6Al–4V alloy. *Journal of Materials Processing Technology*, 211(3), 356–362, 2011.

[113] S. Zhang, J. Li, H. Lv, Tool wear and formation mechanism of white layer when hard milling H13 steel under different cooling/lubrication conditions. *Advances in Mechanical Engineering*, 6, 949308, 2014.

[114] S. Zhang, J. Li, H. Lv, Tool wear and formation mechanism of white layer when hard milling H13 steel under different cooling/lubrication conditions. *Advances in Mechanical Engineering*, 6, 949308, 2014.

[115] S. Sartori, A. Ghiotti, S. Bruschi, Hybrid lubricating/cooling strategies to reduce the tool wear in finishing turning of difficult-to-cut alloys. *Wear*, 376–377, 107–114, 2017.

[116] F. Pusavec, A. Deshpande, S. Yang, R. M'Saoubi, J. Kopac, O. W. Dillon, Sustainable machining of high temperature Nickel alloy–Inconel 718: Part 1–predictive performance models. *Journal of Cleaner Production*, 81, 255–269, 2014.

[117] F. Pusavec, A. Deshpande, S. Yang, R. M'Saoubi, J. Kopac, O. W. Dillon, Sustainable machining of high temperature Nickel alloy–Inconel 718: Part 2–chip breakability and optimization. *Journal of Cleaner Production*, 87, 941–952, 2015.

[118] M. K. Gupta, Q. Song, Z. Liu, M. Sarikaya, M. Jamil, M. Mia, V. Kushvaha, A. K. Singla, Z. Li, Ecological, economical and technological perspectives based sustainability assessment in hybrid-cooling assisted machining of Ti-6Al-4V alloy. *Sustainable Materials and Technologies*, 26, e00218, 2020.

[119] K. K. Gajrani, M. Ravi Sankar, State of the art on micro to nano textured cutting tools, Materials Today*: Proceedings,* 4, 3776–3785, 2017.

[120] P. Koshy, J. Tovey, Performance of electrical discharge textured cutting tools. *CIRP Annals -Manufacturing Technology*, 60, 153–156, 2011.

[121] L. Yunsong, D. Jianxin, Y. Guangyan, C. Hongwei, Z. Jun, Preparation of tungsten disulfide (WS_2) soft-coated nano-textured self-lubricating tool and its cutting performance. *International Journal of Advanced Manufacturing Technology*, 68, 2033–2042, 2013.

[122] T. Sugihara, T. Enomoto, Development of a cutting tool with a nano/micro-textured surface-Improvement of anti-adhesive effect by considering the texture patterns. *Precision Engineering*, 33, 425–429, 2009.

[123] S. Wenlong, D. Jianxin, Z. Hui, Y. Pei, Z. Jun, A. Xing, Performance of a cemented carbide self-lubricating tool embedded with MoS_2 solid lubricants in dry machining, *Journal of Manufacturing Processes*, 13, 8–15, 2011.

[124] S. Devaraj, R. Malkapuram, B. Singaravel, Performance analysis of micro textured cutting insert design parameters on machining of Al-MMC in turning process. *International Journal of Lightweight Materials and Manufacture,* 4, 210–217, 2021.

[125] R. Sharma, S. Pradhan, R. N. Bathe, Design and fabrication of spiral triangular micro texture on chemical vapor deposition coated cutting insert using femtosecond laser machine. *Materials Today: Proceedings,* 28, 1439–1444, 2020.

[126] T. Sugihara, T. Enomoto, Crater and flank wear resistance of cutting tools having micro textured surfaces. *Precision Engineering,* 37, 888–896, 2013.

[127] P. Ranjan, S. S. Hiremath, Role of textured tool in improving machining performance: A review. *Journal of Manufacturing Processes,* 43, 47–73, 2019.

[128] J. Xie, M. J. Luo, K. K. Wu, L. F. Yang, D. H. Li, Experimental study on cutting temperature and cutting force in dry turning of titanium alloy using a non-coated

micro grooved tool. *International Journal of Machine Tools and Manufacture,* 73, 25–36, 2013.

[129] N. Kawasegi, H. Sugimori, H. Morimoto, N. Morita, I. Hori, Development of cutting tools with microscale and nanoscale textures to improve frictional behavior. *Precision Engineering,* 33, 248–254, 2009.

[130] R. Sasi, S. Kanmani Subbu, I. A. Palani, Performance of laser surface textured high speed steel cutting tool in machining of Al7075-T6 aerospace alloy. *Surface and Coatings Technology* 313, 337–346, 2017.

[131] A. Fatima, P. T. Mativenga, A comparative study on cutting performance of rake-flank face structured cutting tool in orthogonal cutting of AISI/SAE 4140. *International Journal of Advanced Manufacturing Technology*, 78, 2097–2106, 2015.

[132] J. Sun, Y. Zhou, J. Deng, J. Zhao, Effect of hybrid texture combining micro-pits and micro-grooves on cutting performance of WC/Co-based tools. *International Journal of Advanced Manufacturing Technology*, 86, 2016, 3383–3394.

[133] W. Chang, J. Sun, X. Luo, J. M. Ritchie, C. Mack, Investigation of microstructured milling tool for deferring tool wear. *Wear,* 271, 2433–2437, 2011.

[134] J. Kümmel, D. Braun, J. Gibmeier, J. Schneider, C. Greiner, V. Schulze, A. Wanner, Study on micro texturing of uncoated cemented carbide cutting tools for wear improvement and built-up edge stabilisation. *Journal of Materials Processing Technology,* 215, 62–70, 2015.

[135] Y. Liu, J. Deng, F. Wu, R. Duan, X. Zhang, Y. Hou, Wear resistance of carbide tools with textured flank-face in dry cutting of green alumina ceramics. *Wear*, 372–373, 91–103, 2017.

[136] D. Arulkirubakaran, V. Senthilkumar, Performance of TiN and TiAlN coated micro grooved tools during machining of Ti–6Al–4V alloy. *International Journal of Refractory Metals and Hard Materials,* 62, 47–57, 2017.

[137] S. Jesudass Thomas, K. Kalaichelvan, Comparative study of the effect of surface texturing on cutting tool in dry cutting. *Materials and Manufacturing Processes,* 33, 683–694, 2018.

[138] W. Ze, D. Jianxin, C. Yang, X. Youqiang, Z. Jun, Performance of the self-lubricating textured tools in dry cutting of Ti–6Al–4V. *International Journal of Advanced Manufacturing Technology,* 62, 943–951, 2012.

[139] J. Deng, Y. Lian, Z. Wu, Y. Xing, Performance of femtosecond laser-textured cutting tools deposited with WS_2 solid lubricant coatings. *Surface and Coatings Technology*, 222, 135–143, 2013.

[140] S. Wenlong, D. Jianxin, W. Ze, Z. Hui, Y. Pei, Z. Jun, A. Xing, Cutting performance of cemented-carbides-based self-lubricated tool embedded with different solid lubricants. *International Journal of Advanced Manufacturing Technology,* 52, 477–485. 2011.

[141] Y. Feng, J. Zhang, L. Wang, W. Zhang, Y. Tian, X. Kong, Fabrication techniques and cutting performance of micro-textured self-lubricating ceramic cutting tools by in situ forming of Al_2O_3–TiC. *International Journal of Refractory Metals and Hard Materials,* 68, 121–129, 2017.

[142] K. Orra, S. K. Choudhury, Tribological aspects of various geometrically shaped micro-textures on cutting insert to improve tool life in hard turning process. *Journal of Manufacturing Processes,* 31, 502–513, 2018.

[143] A. Farooqi, N. B Yusoff, Green manufacturing – Textured novel cutting tool for sustainable machining: *A review. Applied Mechanics and Materials*, 899, 135–143, 2020.

[144] L. Fan, Z. Deng, X. Gao, Y. He, Cutting performance of micro-textured PCBN tool. Nanotechnology and Precision Engineering, 4, 023004, 2021.

[145] Q. Li, C. Pan, Y. Jiao, K. Hu, Investigation on cutting performance of micro-textured cutting tools. *Micromachines,* 10, 352, 2019.

[146] M. S. Sawant, N.K. Jain, I. A. Palani, Influence of dimple and spot-texturing of HSS cutting tool on machining of Ti-6Al-4V. *Journal of Materials Processing Technology,* 261, 1–11, 2018.

[147] J. Ma, N. H. Duong, S. Chang, Y. Lian, J. Deng, S. Lei, Assessment of microgrooved cutting tool in dry machining of AISI 1045 steel. *Journal of Manufacturing Science and Engineering,* 137, 031001, 2015.

[148] W. Song, Z. Wang, S. Wang, K. Zhou, Z. Guo, Experimental study on the cutting temperature of textured carbide tool embedded with graphite. *International Journal of Advanced Manufacturing Technology,* 93, 3419–3427, 2017.

[149] Z. Fang and T. Obikawa, Cooling performance of micro-texture at the tool flank face under high pressure jet coolant assistance. *Precision Engineering* 49 (2017) 41–51.

[150] Y. Su, Z. Li, L. Li, J. Wang, H. Gao, G. Wang, Cutting performance of micro-textured polycrystalline diamond tool in dry cutting. *Journal of Manufacturing Processes,* 27, 1–7, 2017.

[151] K. K. Gajrani, R. P. K. Reddy, M. R. Sankar, Tribo-mechanical and surface morphological comparison of untextured, mechanical micro-textured (MμT), and coated-MμT cutting tools during machining. Proceedings of the Institution of Mechanical Engineers, Part J: *Journal of Engineering Tribology,* 233(1), 95–111, 2019.

[152] K. K. Gajrani, R. P. K. Reddy, M. R. Sankar, Experimental comparative study of conventional, micro-textured and coated micro-textured tools during machining of hardened AISI 1040 alloy steel. *International Journal of Machining and machinability of Materials*, 18, 522–539, 2016.

[153] J. Xie, M. J. Luo, K. K. Wu, L. F. Yang, D. H. Li, Experimental study on cutting temperature and cutting force in dry turning of titanium alloy using a non-coated micro grooved tool. *International Journal of Machine Tools and Manufacture* 73, 25–36, 2013.

[154] D. Vasumathy, A. Meena, Influence of micro scale textured tools on tribological properties at tool-chip interface in turning AISI 316 austenitic stainless steel. *Wear,* 376–377, 1747–1758, 2017.

[155] S. Dinesh, V. Senthilkumar, P. Asokan, Experimental studies on the cryogenic machining of biodegradable ZK60 Mg alloy using micro-textured tools. *Materials and Manufacturing Processes*, 32, 979–987, 2017.

[156] N. Li, Y. Chen, D. Kong, S. Tan, Experimental investigation with respect to the performance of deep submillimeter-scaled textured tools in dry turning titanium alloy Ti–6Al–4V. *Applied Surface Science*, 403, 187–199, 2017.

[157] Y. Xing, J. Deng, J. Zhao, G. Zhang, K. Zhang, Cutting performance and wear mechanism of nanoscale and microscale textured Al_2O_3/TiC ceramic tools in dry cutting of hardened steel, *International Journal of Refractory Metals and Hard Materials,* 43, 46–58, 2014.

[158] S. Niketh, G. L. Samuel, Surface texturing for tribology enhancement and its application on drill tool for the sustainable machining of titanium alloy. *Journal of Cleaner Production,* 167, 253–270, 2018.

2 Dry and MQL Assisted Hard Machining

2.1 INTRODUCTION

Currently, technological advancements are accelerating faster than ever before, and the demand for production and ecological efficiency has substantially increased. In today's industrial world, sustainable manufacturing is considered to be the most recent trend. This provides both economic and ecological benefits. Due to their ecological, economic and technological advantages, sustainable technologies are gaining recognition in the machining sector. Today, technological advancements are occurring faster, and the need for higher efficiency in sustainable production and the environment has grown.

Minimum quantity lubrication (MQL) is one method that permits highly pressured cutting fluid to be applied to the machining zone as tiny oil particles. In comparison to the conventional cooling systems and dry machining, the MQL strategy can be favoured because of the prevention and/or reduction of environmental contamination, high energy consumption and employee health issues. It is worth pointing out that the pursuit of sustainable manufacturing is driving the adoption of cutting-edge new technology, mostly for hard materials. In precise turning, tool wear is a natural growth area for research. In order to highlight the trends and visualize the context, recent literature on sustainable machining is presented here.

The machining parameters, such as cutting speed, are the most impactful on the productivity of the machining process. Although the generation of cutting temperature is mostly dependent on the cutting speed for a given combination of cutting tool and workpiece, there are always constraints on the cutting speed to be used. Green machining, notably MQL, was devised as a compromise between the advantages and disadvantages of dry cutting and machining with abundant soluble oil, when ecological awareness rules and regulations were enacted [1]. There is a need to look at newer cooling lubrication methods to improve sustainability in the machining process and produce a cleaner and greener environment.

A larger emphasis has been put on affordable and eco-friendly MQL approaches. Sustainability refers to a comprehensive participative strategy that encompasses ecological, economic and social implications and has an undeviating effect not only on actual production processes but also on the entire manufacturing industry. In the recent competitive global climate, the reality that governments prioritize economic growth over sustainability has led to alarming levels of environmental contamination. Sustainable manufacturing also implies that organizations can secure the safety of their future output and perform the present production operations with less environmental impact. Therefore, developing sustainable manufacturing has become a priority for both organizations and society [2]. Throughout the past two decades, explorations of the MQL strategy have put forward the qualitative and quantitative correlations between the workpiece material, tool material, cutting fluid, cutting tooling wear and measured surface roughness [3]. Researchers are working to solve the problem of ecologically friendly machining of hard materials. For example, He et al. [4] pointed towards the objective of optimization considering sustainable development and greener and cleaner production. As suggested, multiple-objective MQL optimization should be added in the control of oil mist. Further, it was addressed that research and development of MQL systems should aim towards the establishment of an intelligent MQL method that can be coupled to the machine tool and adapted to gain multi-objective process parameter optimization.

The aim of sustainability in the machining process is to produce a high quality product with a larger degree of precision and surface quality, the lowest energy consumption and environmental impact, and at a reasonable cost. The difficulties and obstacles related to dry cutting can be summed up as follows [5]:

- Deficiency of any cutting fluids produces a strict concentration of heat at the cutting tool–workpiece contact zone resulting in degradation of the tool hardness.
- Exposure to heat has a significant influence on the workpiece during the machining process. This exposure to heat has a harmful impact on the surface topography and subsurface layer features of the machined parts.
- The difficulty of dry cutting grows as the machining indices of the workpiece increase, during the cutting of difficult-to-cut materials.

Chandel et al. [6] evaluated research innovations in the engineering of machining processes guided to further improve sustainability. The authors provided a comprehensive examination of the importance of sustainable machining technology in achieving and executing sustainable manufacturing goals. Kui et al. [7] described that during the machining process a significant portion of the manufacturing industry involves machine parameters, lubricants and the environment. As a result, emphasis has been laid on environmental pollution and health and safety concerns. It turns out that manufacturing organizations were compelled to apply machining techniques which are environmentally friendly. Consequently, MQL has gained considerable attention over the years because of its capacity as a cooling delivery solution to augment lubrication performance and heat transfer for machining processes. Singh et al. [8] reported that the recent conditions necessitated for sustainable machining comprise improved utilization of man, machine and surroundings. Furthermore, a sustainable machining process is the solution to developing a cleaner environment. It was summarized that while placing MQL nozzles away from the machining zone, lubricant droplets cannot support the machining zone's cooling and lubrication action. Zhu et al. [9] noted that MQL is an eco-friendly technology that is useful for achieving sustainability in machining. The nozzle distance plays an important role in controlling the MQL spray. Understanding of the orientation setup of nozzle and equipment development in MQL based milling processes (for external MQL) has been achieved. The results have indicated that over 90% of the rebounded droplets have the capability to re-adhere to the surface of the tool. Zhu et al. [10] explored the penetration MQL performance with various spray factor and capillary size combinations. The results revealed that irrespective of penetration distance and forms, high air flow rate, greater capillary size and small droplet size yielded the shortest penetration time. Sohrabpoor et al. [11] investigated experimentally the impact of different lubrication and process parameters on tooling wear and surface roughness in the turning process of AISI 4340 steel. The minimal amount of lubricant ensures a minimum surface roughness and less tool wear because of a decrease in cutting temperature and enhanced chip flushing. Fernando et al. [12] summarized that the manufacturing industry consumes 40% and 25% of the world's resources and energy respectively. Machining processes are energy-intensive production processes, contributing significantly to the environmental footprint. Özbek and Saruhan [13] researched and observed that as well as affecting tool service life and employee health, the consequences of surface roughness are related to product quality. This has led to development in the usage of the environmentally friendly MQL technology, which is now a competitor to dry and flood machining.

Boubekri and Shaikh [14] reported that MQL serves as a substitute for the flood cooling process by minimizing the cutting fluid volume applied in the machining procedure – but not without significant health concerns. It was indicated that MQL use produces an important amount of mist compared to the flood cooling process. Abbas et al. [15] reported that due to the constantly increasing global demand for high-quality and sustainable products, the scientific and manufacturing communities are currently working on sustainable machining assessment and design, with a special emphasis on developing innovative and advanced metal cutting technology solutions. Paturi and Reddy [16] summarized that dry machining is performed without cutting fluids, therefore this environment is more environmentally friendly than the wet machining process. In this text, MQL is presented as a cleaner substitute for traditional machining as a result of the shortcomings of traditional machining techniques. MQL is considered to be a green machining process that has the potential to improve the machining indices of hard-to-cut materials. Further, to promote cleaner sustainable machining, various sectors are alternating traditional machining with MQL. In their review, Kawade and Bokade [17] noted that the selection of an appropriate cooling method for every machining process is one of the most significant challenges for manufacturers, as it can minimize tool wear and difficulties related to cutting temperature. Further, augmentation in temperature at the workpiece–tool contact accelerates cutting tool wear in many machining applications. Katna et al. [18] reported that massive quantities of cutting fluid are used by manufacturing organizations. Cutting fluids are important to the industrial sector as they limit the effects of heat, reduce friction, offer lubrication, aid in the removal of chips and swarf, decrease built-up-edge (BUE) formation on cutting tools and enhance machining performance and tool service life. Vegetable oils are an excellent substitute for standard cutting fluids. Bag et al. [19] emphasized that sustainable machining provides environmental friendliness, minimizes power consumption and advances personal safety and health.

2.2 MACHINING PERFORMANCE UNDER DRY ENVIRONMENT

In recent years, researchers have focused on the tool service life induced by attrition in dry precision hard cutting of hard-part steel and have already achieved remarkable outcomes. For example, Goindi and Sarkar [20] noted that machining is a manufacturing process used to accomplish the desired final shape and surface characteristics of a manufactured component. Dry cutting leads to a rise in cutting temperature that makes chips softer. These formed chips stick to the tool and workpiece surfaces, resulting in tool chipping, workpiece surface damage and chip entanglement. In the machining process, a great deal of heat is generated, and the removal of heat needs the use of appropriate fluids that are a significant source of waste and ecological harm. Sreejith and Ngoi [21] pointed out that, due to environmental concerns, dry or green machining is growing in popularity. The report confirmed that 16–20% of manufacturing expenses are attributable to the coolants and lubricants used in machining; hence, the excessive usage of these fluids should be limited. Derflinger et al. [22] noted that, in recent years, dry machining has gained popularity because of environmental regulations, worker safety concerns and cost savings associated with fluid cutting. Due to economic and environmental concerns, dry machining is a highly attractive process. Dry machining dispenses with the need for cutting coolants. Haapala et al. [23] summarized that sustainable manufacturing necessitates simultaneous assessment of the economic, ecological and social features of the formation and delivery of goods. In addition, sustainable manufacturing is dependent on descriptive metrics, public policy for implementation, and sophisticated decision-making and feedback. Debnath et al. [24] pointed out that dry machining is suited to steels, steel alloys and cast irons for turning, milling and gear cutting. Nevertheless, dry machining has limitations in machining processes that generate a great deal of heat, such as drilling. Turning of hard materials usually presents poor machinability. In this context, Mondal et al. [25] turned hardened (43 HRC) 16MnCrS5 steel and further investigated to evaluate cutting performance with plain and TiC coated carbide inserts (wide-groove-type chip breaking) in dry and wet environmental circumstances, with varying cutting speed and feed. Tang et al. [26] explored the machinability of tool wear and mechanisms with a PCBN insert in dry precision hard turning of an AISI D2 tool steel bar. The consequences demonstrated that the workpiece's hardness has a significant effect on flank wear. Figure 2.1 shows that crater type of wear on the rake face raises the mechanical load in the machining zone close to the cutting edge. This causes the temperature to rise, which makes the compressive stress improve.

Abbas et al. [27] assessed the surface quality produced during the turning of AISI 4340 steel with wiper nose and standard round inserts under various cutting conditions. The data revealed that minimum surface roughness was achieved utilizing wiper inserts as compared to traditional tools for the stated range of environments, showing improved performance.

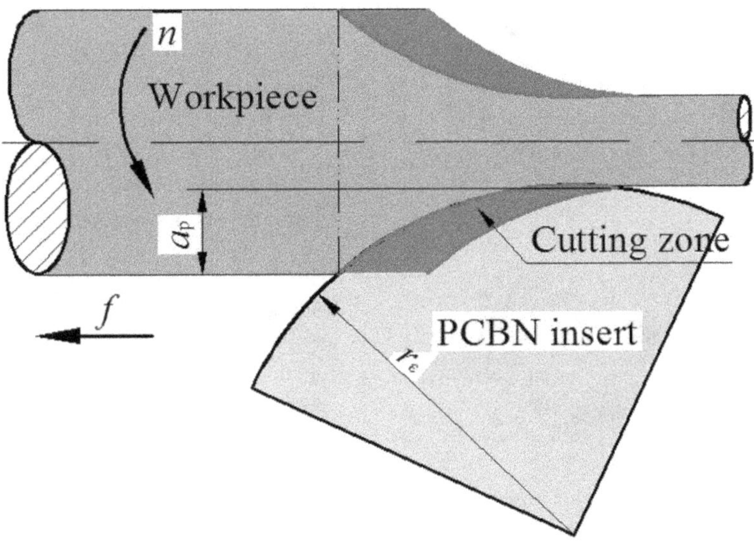

FIGURE 2.1 Diagram of dry hard turning. (From Ref. [26].)

Hard machining tests were carried out by Jomaa et al. [28], investigating the microstructure and residual stresses and changes caused by the cutting of induction AISI 4340 heat-treated steel (58–60 HRC). It was found that surface residual stresses tend to become tensile as cutting speed increases, while compressive stresses grow as feed rate increases. Erkan Öztürk [29] observed the impacts of sectioned micro-textures oriented (circular, square and triangle) on the direction of chip flow on the rake face and attempted on the flank face to lower the nose radius of the tool for hard-part turning of AISI 4140 with composite coated and uncoated inserts. This finite element method (FEM) and statistical study found that texturing inserts did not affect AISI 4140 machining performance. Dikshit et al. [30] examined the influence of biocompatible TiAlN coated and uncoated carbide inserts, and the impact of feed, rotating speed and depth of cut on the surface roughness of M2 tool steel (64 HRC). It was determined that a TiAlN coated carbide insert provides a smoother surface than an untreated carbide insert. Orra and Choudhury [31] explored the effects of horizontal, vertical and elliptical micro-textures infused with MoS_2 dry lubricant on the rake face surface of hard turning tools. The research found that vertically textured cutting inserts were more successful, with a maximum coefficient of friction reduction of 11.9%. Figure 2.2 illustrates the 3D profile surface picture with depth and sectional profile graphs evaluated in the textured channel by a surface profilometer. It demonstrates how the height of the grooves is the height over the tool rake surface.

The first study performed by Saikaew et al. [32] presents the concurrent impact of the cutting speed and the tool materials, and the interface between the speed and the tool materials Al_2O_3+TiC (ceramic composite tools) and TiN+AlCrN (coated cemented carbide) on the surface roughness and cutting tool wear during the turning process of AISI 4140 steel under dry conditions. As suggested by the results, Al_2O_3+

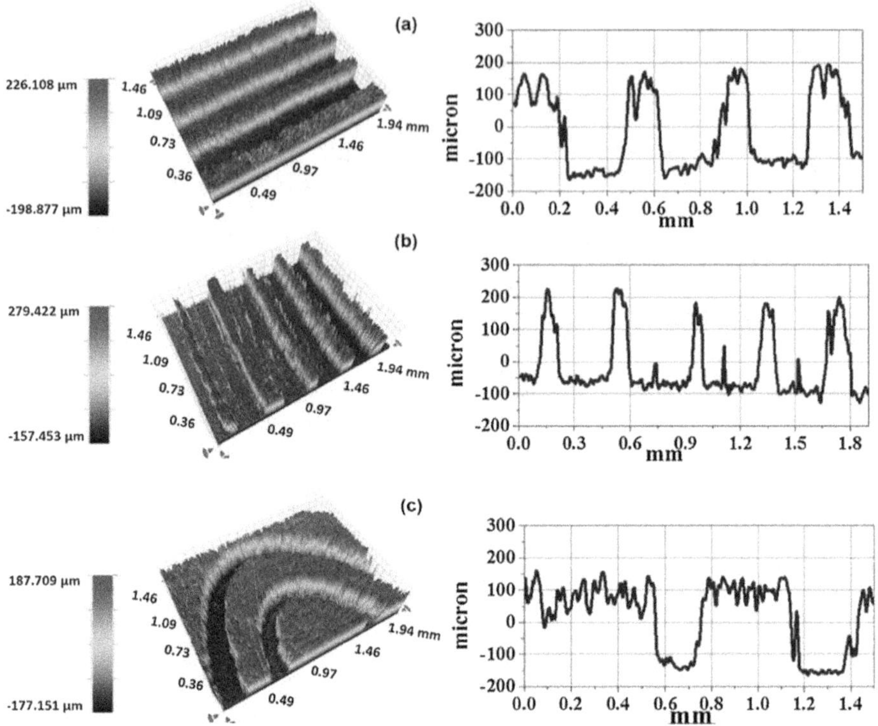

FIGURE 2.2 Picture of micro-textures on tool showing main chip flow direction and cutting edge after machining (a) vertical micro-textured cutting insert, (b) horizontal micro-textured cutting insert and (c) elliptical micro-textured cutting insert. (From Ref. [31].)

TiC tools may be more appropriate for hard turning of extremely hard ferrous materials than TiN+AlCrN tools. Figure 2.3 presents the experimental process for the hard-part turning analysis of AISI 4140 steel.

Asiltürk et al. [33] determined the impact of cutting speed, feed rate and depth of cut on acoustic emissions, vibrations and surface roughness during the dry turning process. It was deduced that the mixture of feed rate and cutting speed must be adjusted in order to prevent excessive vibrations when cutting hardened steels. Nonetheless, the depth of incision has a significantly more consistent effect on vibrations. The feed rate controls the variation in machined surface roughness. Due to the combined impact of cutting settings and the difficulty of the machining process, the evaluation of acoustic emissions is extremely difficult. According to Das et al. [34] it was recognized that machining processes under dry cutting conditions are environmentally friendly and techno-economically feasible for enhancing sustainability. Further, to improve sustainability performance, it was suggested that power consumption, cutting temperature, machining cost and waste management under dry hard turning processes may be considered. Ishfaq suggested that [35] dry machining is favourable from an environmental standpoint, however, it results in inferior product surface

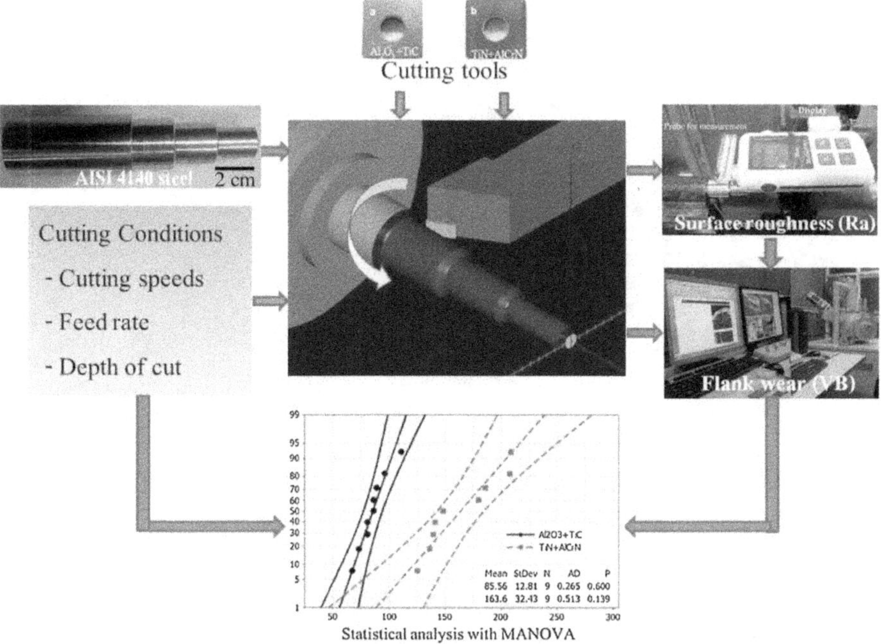

FIGURE 2.3 Experimental process for hard turning analysis. (From Ref. [32].)

integrity and greater tooling costs. Rath et al. [36] conducted dry turning of D3 steel for better understanding of the mechanism with a coated mixed ceramic tool insert. At a minimum cutting speed, the main factor that controls the tool flank wear is the feed rate. It was divulged from the chip morphology that wear behaviour and the friction coefficient control the thickness and width of the chip.

2.3 MACHINING PERFORMANCE UNDER MQL ENVIRONMENT

Mawandiya et al. [37] turned AISI 4340 steel with a carbide insert in dry, MQL and minimum quantity solid lubrication (MQSL) cooling/lubrication settings to determine the viability of the cooling or lubrication technology. Figure 2.4 shows that more energy is consumed under dry machining because of more tool wear, resulting in higher carbon emission levels.

Bhadoria et al. [38] noted that for machining, the MQL approach has gained attractiveness because of its ability to reduce forces and tool wear. MQL also provides economic and environmental advantages in machining by controlling the quantity of coolant-lubricant fluid used during milling. The authors summarized that the use of MQL would augment the process productivity in terms of parts per hour. Vadaliya et al. [39] pointed out that in the industrial sector, metal cutting is a promising approach for achieving appropriate surface polish and precision on the finished workpiece. Several tactics, including the application of cutting fluids, coated carbide tools, surface texturing on the rake face etc., are now being applied to improve

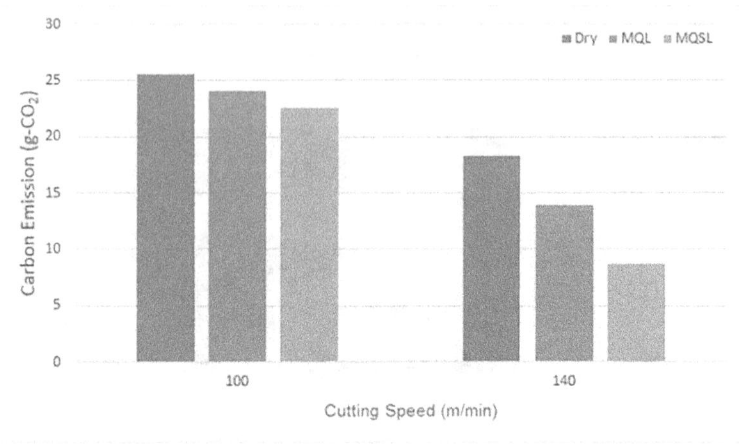

FIGURE 2.4 Carbon emissions generated under dry, MQL and MQSL machining environments. (From Ref. [37].)

surface finish. Of the known methods, the application of cutting fluid through metal machining is the most promising for enhancing surface integrity and reducing total machining costs. Rafighi et al. [40] accomplished hard tuning on AISI D2 tool steel (60 HRC) under dry cutting conditions. The results revealed that nose radius has a significant outcome on surface roughness. Özbek et al. [41] summarized that dry cutting is favoured, since there is no cutting fluid that must be discarded after processing. Furthermore, the nonexistence of cutting fluid renders it safe for employee health. Because of the higher temperatures in the cutting zone, however, tool wear increases, which has a detrimental impact on surface roughness and tool service life. In a research examination of Das et al. [42], the MQL technique was used to examine the influence of cutting fluids (compressed air, water-soluble coolant type and Al_2O_3-based nanofluid) on different machining forces, chip thickness and tool flank wear. The authors confirmed that depth of cut is the most important cutting factor in cutting force. Analysis of chip morphology (see Figure 2.5) revealed white and black bands known as ridges and feed marks respectively. There is production of higher temperature where ridges are formed.

Singh et al. [43] summarized that the application of sustainable manufacturing techniques effectively contributes to the preservation of the environment. Modern cooling tactics executed in the manufacturing industry have presented solutions that are conducive to economic expansion and environmental sustainability. Muaz et al. [44] summarized that MQL, also known as near-dry machining (NDM), recently earned an enormous amount of importance for scientists. This approach significantly decreases friction, cutting tool wear and cutting forces compared to conventional machining. MQL's overall performance is determined by the cutting fluid selected. The solid lubricants present in MQL produced superior results in reducing cutting forces, extending tool service life and improving machined surface quality. A lot of extensive experimental studies have been conducted to decrease cutting

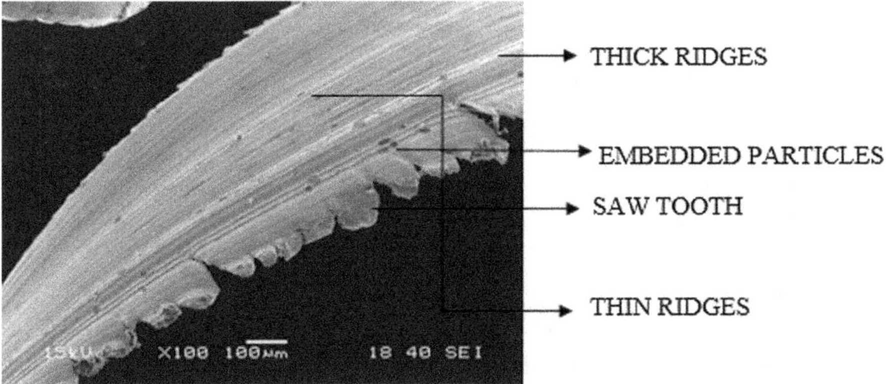

FIGURE 2.5 SEM picture of chips using compressed air. (From Ref. [42].)

fluid consumption and provide cleaner cutting fluids to enhance turning process sustainability. For example, Singh et al. [45] summarized that during mass production, dry cutting is confined to a specific range of cutting factors and materials. Thus, its utility can be increased by coating tools and optimizing process settings. MQL can be applied successfully to the restrictive conditions of dry machining. In particular, MQL is utilized for machining the sophisticated materials used in aerospace and biomedical applications, as dry machining these materials would not be cost-effective because of their high temperature, friction and wear. Figure 2.6 presents the diverse approaches to sustainable manufacturing for improved resource utilization, while the different costs acquired by traditional flood cooling processes are expressed.

Selvam and Sivram [46] analyzed the surface roughness attained during the turning of AISI 4340 steel with flooded, near-dry and dry conditions, using three coated carbide cutting tools. The results showed that NDM provides comparable product surface quality to flooding machining. Mazumder et al. [47] turned AISI 4340 steel at operating conditions in the range of speed 210–350 m/min, feed rate 0.08–0.2 mm/rev and depth of cut 0.5–1.5 mm. The maximum stress was found to be 1311 MPa at the cutting speed 350 m/min, depth of cut 1.5 mm and feed rate 0.08 mm/min. Gunjal et al. [48] summarized that cutting fluids are widely utilized in industry to tackle the difficulties of chip formation, surface roughness, cutting temperature, cutting forces, tool failure and tool wear etc. The use of cutting fluids reduces the harmful consequences of temperature and friction on the machining condition. Sharma and Sidhu [49] observed the impacts of dry machining and NDM on AISI D2 steel at various cutting speed–feed rate combinations using vegetable oil (environmentally friendly) as a lubricant. Chetan et al. [50] summarized that MQL has evolved as a better approach that reduces biological and ecological damage by the widespread application of metalworking fluids. The outcomes of machining with the smallest volume of fluid were similar or superior to traditional flood cooling processes. In the MQL scheme, chips might be collected directly in a dry state, saving time and money on consequent chemical processing for recycling. Figure 2.7 highlights the advantages and limitations associated with sustainable techniques.

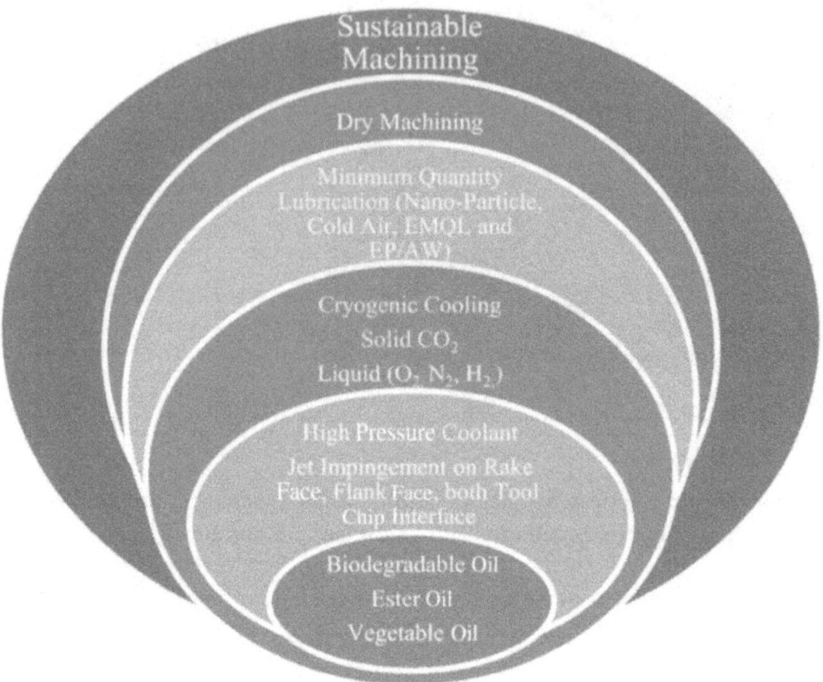

FIGURE 2.6 Extensive sphere of sustainable machining. (From Ref. [45].)

2.4 SUSTAINABILITY ASSESSMENT

It is a truth that energy usage may be reduced to reduce CO_2 emissions. Reducing electrical energy usage can lower energy costs, but it is not sufficient to enhance the overall efficiency of a process. As a result, it is necessary to study the impact of resource-based energy consumption on the sustainability of processes. The different characteristics of sustainable machining approaches [50] are presented in Figure 2.8.

Few studies have looked into hardened steel machining. Sustainable lubrication techniques to lessen the environmental impact of cutting fluids predominate in the literature on sustainable metal cutting.

The researchers Chakule et al. [51] explained key features of the grinding process, including surface roughness and force of cutting. A responsive surface approach (RSM) was applied to horizontal surface grinding machines in the studies. Pušavec et al. [52] highlighted that financial crisis, worldwide rivalry, stronger environmental legislation and supply-chain demand are exerting growing strain on businesses. To address these difficulties, a sustainable development trend must be ensured on all levels/fields, including manufacturing processes. This article encourages sustainable production by enhancing machining technology using a cryogenic machining alternative. In the first section of this study, production challenges in terms of society, the economy and the environment in light of machining technology are examined. Specifically, the study

Sustainable techniques	Advantages	Drawbacks
▪Cryogenic treatment	▪Improves the wear resistance of cutting tool	▪Not a suitable method for long duration of machining
▪Cryogenic cooling	▪No chance of any environmental deterioration as liquid nitrogen is used as a coolant	▪The production cost of cryogen is very high in comparison to other processes
▪MQL	▪Relatively less costly method in comparison to other processes	▪Mist formation is still a major drawback
▪Solid lubrication	▪It is a more sustainable approach as compared to wet cooling	▪Production of the solid lubricants is a costly process
▪Alternative cutting fluids	▪Alternative cutting fluids such as vegetable oil and ionic liquids are totally biodegradable and renewable	▪Vegetable oil has poor thermal stability

FIGURE 2.7 Advantages and limitations associated with sustainable techniques. (From Ref. [50].)

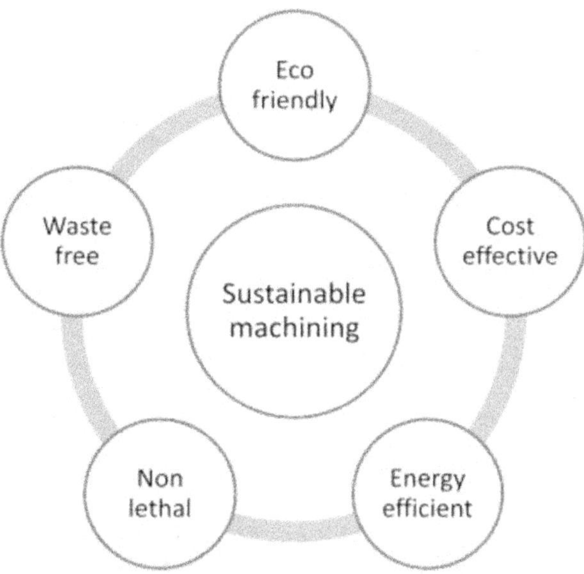

FIGURE 2.8 Features of sustainable machining. (From Ref. [50].)

focuses on the fundamental pillars of sustainability, such as reduced energy usage, increased tool life, enhanced final product functionality through improved surface integrity, etc. In addition to analyzing the performance of the machining process, a study of the total cost of the process was also conducted, revealing that cryogenic machining can significantly reduce costs compared to traditional machining. Iqbal et al. [53] pointed out that tool life and its performance criteria have been crucial for monitoring sustainability measures such as process cost, energy usage and work quality. According to the review results, the tool service life criterion has a very significant effect on all of the sustainability indicators. From the perspective of sustainability, Ghamdi and Iqbal [54] investigated experimentally and compared conventional and high-speed machining. Using the concepts of green and sustainable mechanisms, the reported work by researchers is expected to find applications in the metal cutting industry. Khan et al. [55] summarized that the use of two nozzles may maintain 150% more aggressive cutting speed than standard dry machining without sacrificing tool life. Industry and academia have increasingly been looking for innovative lubricating solutions that not only increase machining efficiency but also decrease resource-based consumption of energy and carbon footprints. Figure 2.9 represents the environmental and machining distinctiveness of the hard milling process of EN31.

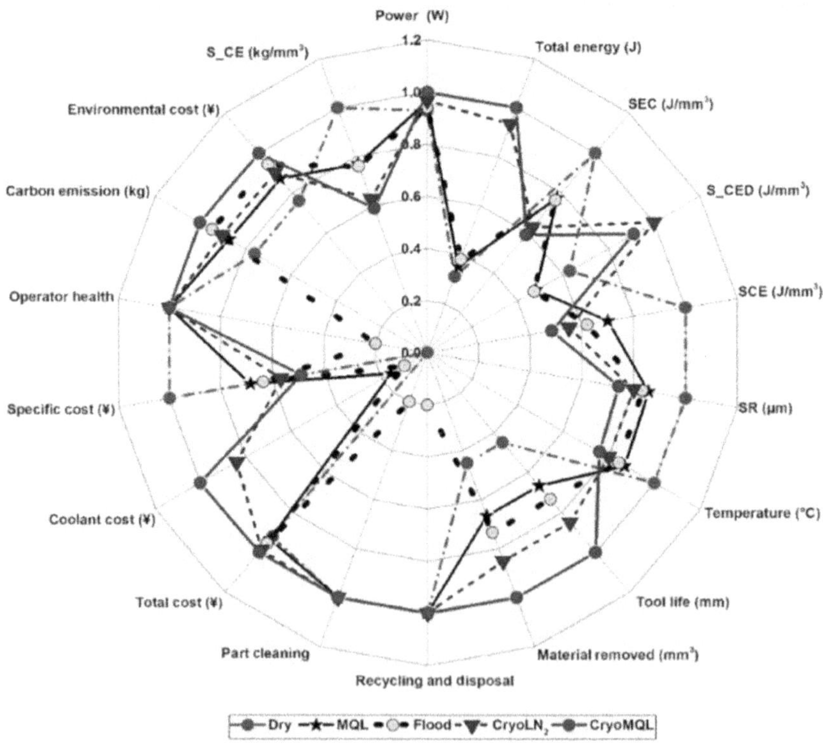

FIGURE 2.9 Environmental and machining characteristics of the hard milling process of EN31. (From Ref. [55].)

Hybrid cryogenic MQL (CryoMQL) emitted the greatest amount of carbon emissions and caused the greatest expense to the environment. To enhance the effectiveness of the CryoMQL assisted end milling process of AISI 52100 steel, these two concerns must be resolved immediately. From Figure 2.9, it is evident that the overall sustainability of CryoMQL is very high. Sustainability assessment refers to energy consumption, environmental impact, safety, waste management, personal health and cost. Şahinoğlu [56] explored the impacts of various operation variables and cooling mechanisms on machinability (energy consumption, power consumption, vibration and surface roughness) under dry, CO_2 and coolant conditions. It was revealed that power consumption is increased by increasing processing parameters. Khanna et al. [57] focused on a comprehensive literature study of life cycle assessment (LCA) analyses performed on machining operations, comparing cutting fluid methods. In addition, the paper gives a comprehensive explanation of the steps required for LCA analysis. Sarikaya et al. [58] highlighted that there are inconsistencies between environmentally responsible and high efficiency machining in machining operations. This suggested that there is a substantial opportunity for sustainable manufacturing to simultaneously boost efficiency and preserve ecological equilibrium. In order to limit resource consumption, it is now essential to adopt efficient procedures. Sihag et al. [59] summarized that, in recent years, there has been an increase in energy costs, environmental concerns, government regulations and consumer awareness of sustainability. It has prompted manufacturing businesses to replace their typical financial machining goals with sustainable plans that include environmental and social factors in addition to economic considerations. A sustainability evaluation index for machine tools was presented, which is intended to help manufacturers and users examine the sustainability performance of machine tools, further to provide decision makers with clear information and support the switch towards greener machine tools. Manufacturing businesses are under direct and indirect pressure to implement sustainable machining methods due to carbon price regulations, government law and customer demand. Sutherland et al. [60] summarized that the manufacturing industry influences all three facets of sustainability: the economy, the environment and society. They examine the social impacts recognised by social indicators, frameworks and principles at the national level. Several stakeholder groups with associated social requirements are considered when analyzing the influence of manufacturing on social performance.

Olufayo et al. [61] emphasized that sustainable manufacturing in Industry 4.0 is an emerging research subject due to the need for higher production resulting from a growing population and the progress of digitalization. This research is aimed at determining the optimal machining operation, and it is usual practice to conduct research in this area that compares the costs of dry and wet (lubricated) machining procedures.

Jayal [62] looked at how the manufacturing sector, which is central to industrial economies, must be made sustainable to preserve the high standard of living obtained by industrialized nations and to allow developing societies to achieve the same quality of living sustainably. Moreover, the attempt to increase sustainability must provide advantages at all essential levels: environmental, economic and social. Jayal [63] summarized how LCA seeks to assess the overall environmental and economic impact – in terms of material, carbon footprint and energy consumption, etc. – across the product life cycle, from the extraction of raw materials through to its eventual disposal. Despite its tremendous promise, the actual application of LCA as a tool for evaluating design choices for consumer items has fallen short of its potential application because it can become a cumbersome, overly detailed process. The purpose of work by Rotella et al. [64] is to analyze the sustainability performance of machining AISI 52100 through-hardening steel, taking into account the influence of the material removal process in its many facets. Tests were conducted using a chamfered CBN tool and dry and cryogenic cutting conditions. In order to analyze the impacts of extreme in-process cooling on the machined surface, mechanical power, tool wear, cutting force, surface roughness, residual stress and white layer thickness analyses were examined. Khan et al. [65] noted that, in terms of material removal, the Xcel inserts (changed tool geometry) have surpassed conventional and wiper inserts, and the workpiece surface roughness is comparable or superior. In order to determine tool life, wear processes, material removal, power consumption, workpiece surface roughness, microhardness and microstructure, the output responses were monitored and compared to those of recently published works on AISI D2 steel turning.

Das et al. [66] focused on adopting the concept of "Go Green – Think Green – Act Green", a novel technique of economic evaluation and sustainability assessment in hard turning. A newly developed HSN^2-TiAlxN coated carbide insert for hard turning application promises excellent machinability performances. Celent et al. [67] found that increasing machining performance while simultaneously meeting the requirements of sustainable manufacturing represents a significant issue for the metalworking industry and the scientific community. Using multiple types of cutting fluids or optimizing their use during the machining process is one method for overcoming the aforementioned obstacle. In the metalworking business, it is widely recognized that cutting fluids are significant environmental pollutants, which has prompted researchers to develop novel environmentally friendly methods of cooling and lubricating throughout the machining process. Ebrahimi et al. [68] noted that one attractive alternative to conventional machining is one of the sustainable manufacturing technologies called the hot machining technique, which improves machinability and sustainability by minimizing the machining forces, improving the work surface quality and eliminating cutting fluid application. It is obvious that the cutting fluids and coolant/lubricants used in cutting and abrasive processes contain environmentally harmful or potentially damaging chemical constituents.

Pervaiz et al. [69] mentioned that machining methods are widely used in manufacturing because they can provide tight tolerances and high dimensional perfection at a low cost for mass output. Metal cutting processes are often criticized for their considerable environmental consequences and excessive resource input. Green cutting methods are being investigated and compared by metal cutting researchers. These eco-friendly procedures primarily cover metal cutting tasks, and the industrial sector is required to increase its sustainability. Metal cutting input supply and waste stream techniques are being questioned due to their environmental impact. Tough environmental policies are being developed and implemented to enhance metal cutting sustainability. Environmental laws and regulations are putting pressure on manufacturers to include sustainability into their business strategies in order to increase revenue and growth. Environmental and social concerns, in addition to profitability, are being included in corporate performance. Usluer et al. [70] pointed out that, for producers, measuring sustainability has always been a difficult challenge. Several sets of metrics, indices and frameworks have been developed in the literature to assess the sustainability of the manufacturing industry. One of the key aspects of sustainability is the economic factor. Further, it was emphasized that under higher cutting speeds, the sustainable aspects of the machining process were strengthened more than under lower cutting speeds. To achieve more sustainable machining in the proposed study, the effect of dry, MQL, multi-walled carbon nanotube (MWCNT) NFMQL and MWCNT/MoS_2 HNFMQL cutting conditions on the cutting force, thrust force, cutting temperature, total machining cost and carbon emission in the orthogonal turning of S235JR structural steel was investigated. Figure 2.10 depicts the five significant criteria considered when determining the total cost. In addition to machinability assessment, the measurement of sustainability factors in the machining process in terms of total machining costs and total carbon emissions could provide further insight into optimizing the performance and efficiency of the chip removal process. A sustainable machining process is one that uses fewer resources, is cost-effective, is safe for society and has fewer adverse environmental effects.

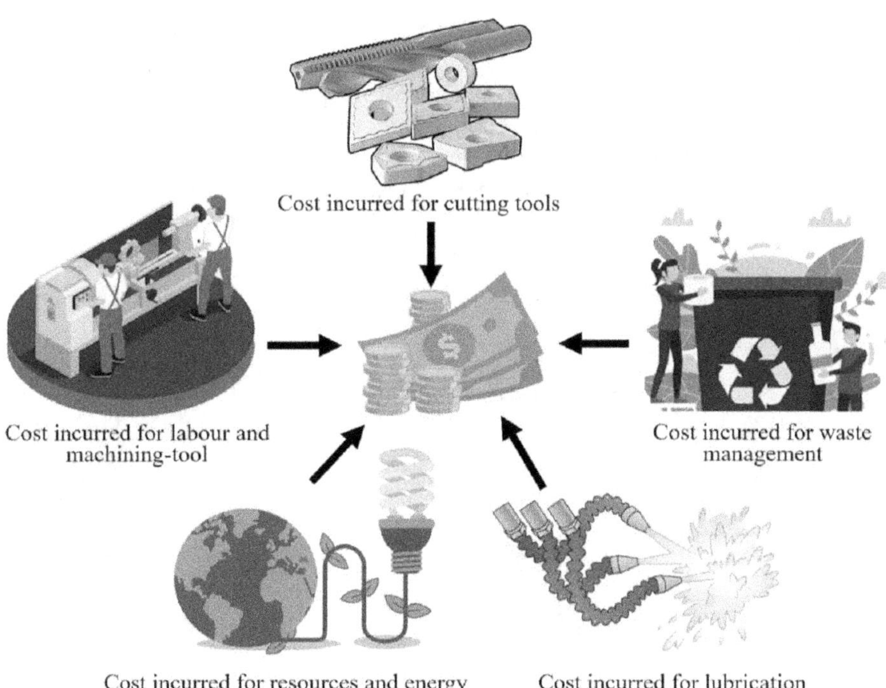

FIGURE 2.10 The key components of overall cost. (From Ref. [70].)

2.5 ADVANCES IN MQL SYSTEMS

Based on a review of the literature, it is evident that the application of hybrid nanofluids is considered as the new machining trend towards sustainability. However, there is a lack of research into sustainability assessment in the context of carbon emission and machining costs using the hybrid nanofluid method. The overall carbon emissions during orthogonal turning of S235JR structural steel under the NFMQL cutting environment were found to be better by up to 60% in comparison with MQL and 37% in comparison with HNFMQL and also in comparison with other cutting environments. The sustainable aspects of the machining process were enhanced more under higher cutting speeds than lower cutting speeds [70].

Zaman and Dhar [71] designed a double jet nozzle for MQL application and investigated experimentally the impact of nozzle angle, size, oil flow rate and air pressure, as well as MQL parameters, on surface roughness and cutting temperature in turning medium carbon steel by coated TiCN tungsten carbide insert. Figure 2.11 presents a diagram of the experimental MQL setup. The parts include compressor, fluid chamber, micro-nozzle and mixing chamber.

Purvis et al. [72] emphasized that, so as to measure how effectively the chip removal procedure performs concerning economics, the environment and society, technical advancements have made the notion of sustainability of paramount importance. In another work, Dash et al. [73] studied the impact of graphene NFMQL machining conditions on AISI D3 steel turning performance. According to the findings, machining with NFMQL provided an effective cooling-lubrication strategy, safer and cleaner manufacturing, and assistance in boosting sustainability. Gajrani [74] et al. used hardened AISI H-13 steel for evaluation of the properties and biodegradability of bio-cutting and mineral oil-based cutting fluids (commercially available) and their hard machining performance (cutting force, feed force, workpiece surface roughness and coefficient of friction). The surface roughness of the workpiece is reduced when

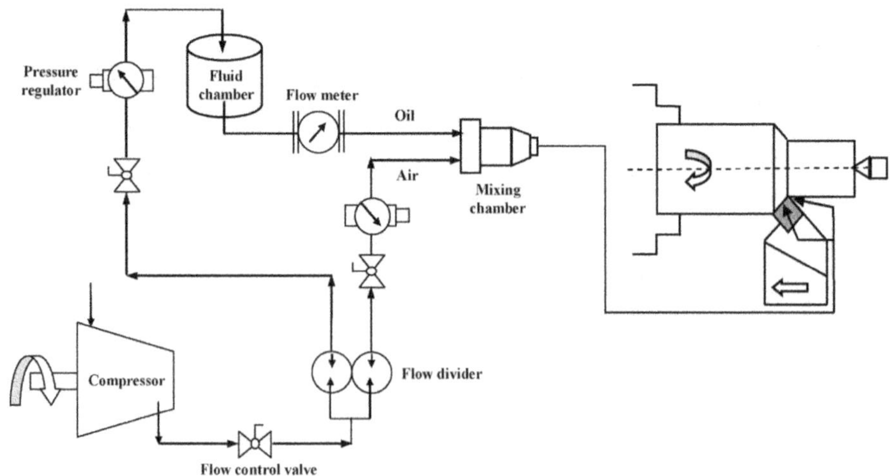

FIGURE 2.11 Schematic picture of the experimental MQL system. (From Ref. [71].)

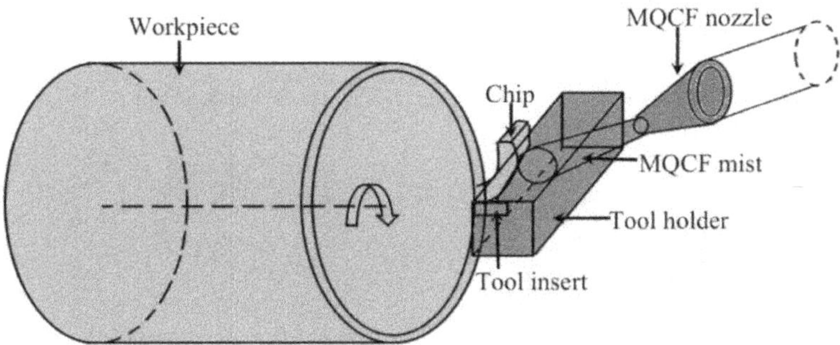

FIGURE 2.12 Schematic of MQCF mist and nozzle direction. (From Ref. [74].)

minimal quantities of cutting fluid are used. Figure 2.12 presents a diagram of the minimum quantity cutting fluid (MQCF) mist and position of nozzle angle.

Khanna et al. [75] turned 15-5 precipitated hardened stainless steel with various cutting fluids, speeds and feed parameters, and this was examined for environmental impacts. Dry cutting uses the most energy. Optimizing cutting settings, reducing particular energy drawn from the network during machining and lowering electrical energy consumption can improve process sustainability and maintain quality. Javid et al. [76] completed a one-step investigation of the sustainability of turning high strength low alloy (HSLA) steel under various machining environments. The authors proposed employing MQL with SiO_2 nanofluids as opposed to MQL base fluid because of decreased energy consumption, less tool wear, improved surface integrity and lower prices. Muthuswamy [77] stated that consumption in face milling can be reduced by 50%, and waste in multiple form deteriorated tools can be eliminated.

Tang et al. [78] proposed that there exists a method in which cutting parameters are optimized in order to establish a balance between process efficiency and carbon emissions [55]. Abbas et al. [79] improved the machining performance with reduced carbon footprint, low machine tool energy consumption and enhanced manufactured products at the lowest cost in response to constant pressure on enterprises to create and adopt sustainable practices. The authors' improved cooling lubrication method has been tested during machining and used as the foundation for sustainability assessment. This evaluation is conducted with regards to surface quality and power consumption, as well as environmental influence, cost of machining, waste management and operator safety and health. A rigorous optimization has been conducted to improve sustainability. In addition, a solution for increased machinability and statistically validated mathematical models of machining responses have been proposed. Kishawy et al. [80] focused on how sustainable machining can be achieved by process optimization, lubrication and cutting conditions. Demand on enterprises to create and adopt sustainable processes has spurred research into machining with low carbon footprint, low machine tool energy consumption and enhanced items at the lowest cost. Over the past few decades, manufacturing energy use has increased dramatically. Pušavec et al. [81] highlighted a life cycle study of the sustainability challenges on the shop floor and concluded that the future of sustainable production will involve the

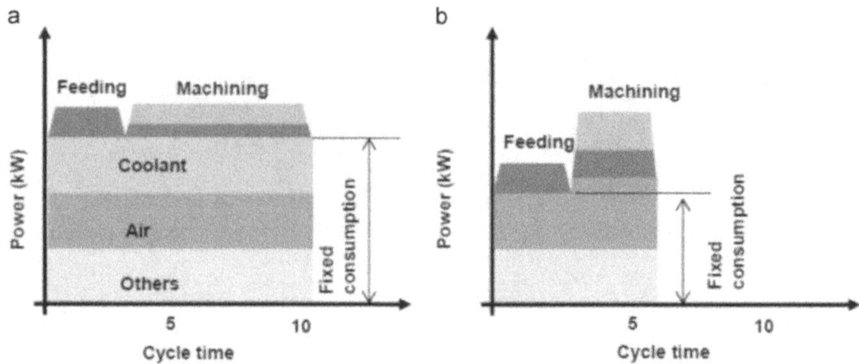

FIGURE 2.13 Diagram of consumption of energy in (a) wet and (b) MQL machining circumstances for an identical process. (From Ref. [83].)

use of cryogenic machining to minimize environmental health risks while improving machining output. In addition, an analysis was conducted to find out the expected effects of cryogenic technology on production costs on machining evaluation. Li et al. [82] noted that future requirements for sustainable development pose a significant challenge and opportunity for conventional manufacturing. Intelligent manufacturing will be able to meet these demands. Evaluations of sustainability play a crucial part in the future development of intelligent manufacturing. Najiha et al. [83] found that the consumption of energy associated with cutting fluid is large. Figure 2.13 illustrates an ideal energy map for a wet and an MQL process.

Das et al. [84] showed that, compared to wet conditions, machining with the nonexistence of cutting fluid is an environmentally friendly, economically viable method for enhancing sustainability. Pusavec et al. [85] presented that, in addition to the central concept of sustainability, cryogenic machining is one of the solutions/alternatives utilized in machining operations to avoid utilizing oil-based emulsions that pose health and environmental risks. Obviously, this is only one solution for the machining process. However, further viable solutions are required for various challenges in machining, manufacturing, etc. Sutherland et al. [86] highlighted that manufacturing enterprises are striving to attain sustainability through changes in processes, products and systems. The results obtained by Rotella et al. [87] indicated that cryogenic cooling has the potential to be used to improve surface integrity for longer product life and greater functional sustainability. Fernando et al. [88] showed that the manufacturing industry consumes 40% and 25% of the world's energy and resources, respectively. Machining is one of the most energy-intensive production processes, contributing significantly to the environmental footprint. Balogun et al. [89] developed a new statistical model and reasoning for calculating machining toolpath direct electrical energy consumption. This model will track energy visibility, process dependence and carbon footprint in the cutting process. This information is essential for evaluating toolpaths and redesigning machine tools to save electricity and reduce carbon footprints. Rajemi et al. [90] mentioned that demand for energy drives carbon dioxide emissions and climate change, and that sustainable manufacturing requires energy reduction. An agency [91] summarized that LCA is the most frequent method used for examining the environmental impact of manufactured items. LCA, as defined by ISO-14040, takes into account the resource use and release consequences of a product throughout its life cycle, from raw material acquisition to manufacturing (cradle to gate), usage, end-of-life recovery, and disposal (cradle to grave). Jawahir et al. [92] provided a comprehensive review of the four life cycle stages and three sustainability components of a product. This novel method will assist product designers and manufacturers in analyzing the overall product and improving future product family updates for economic, environmental and social sustainability. Dahmus et al. [93] performed environmental analysis considering material removal, material preparation and cutting fluid preparation. This system-wide perspective enhances machining evaluation. Material removal, according to energy consumption estimates, consumes less energy than machine tool operation. Huang et al. [94] highlighted that the purpose of this research is to build computational machining energy estimation tools throughout a product's early design phase. In the preliminary or embodiment design, the shape and materials of a product are established. At this stage, it is essential for a designer to be able to assess alternative designs and materials based on a variety of factors, including cost, functionality, energy efficiency, etc. The need exists for automated energy consumption estimation tools that can be integrated with CAD/CAM systems. This paper introduces computational techniques for evaluating the energy consumption of machining operations during the early stages of design. Chaurasiya et al. [95] mentioned that the manufacturing industries are the primary contributors to a nation's economic growth, despite their significant negative environmental repercussions, which are mostly caused by diverse process factors. Carou et al. [96] highlighted that LCA is the most commonly employed method for evaluating the environmental implications of manufactured products. It has also been utilized in machining because it permits a holistic method that considers all environmental exchanges of a product or process throughout its life cycle. When an appropriate baseline is established, it enables the comparison of situations in order to identify those that are more environmentally friendly. However, due to the influence of machinability on the process, comparisons between setups that involve, for example, different workpiece materials are of limited use.

Benedicto et al. [97] stated that the objective is the elimination of all cutting fluids. Due to the stringent necessities of machining procedures, dry machining conditions are inapplicable in some instances. Causes include the extreme heat generated by the process, the increase in friction between the tool and the workpiece, and the requirement to remove formed chips. In addition to spurring the development of novel formulations for cutting fluids, the demand for eco-friendly goods also encourages the creation of sustainable products. Pusavec et al. [98] presented that, on the macro level of production, the concept of sustainable development is well-defined and realized, but there is a serious shortage of implementation techniques on the shop floor including machining technology. Llanos et al. [99] proposed a novel cryogenic CO_2 delivery system that enables a stable and controlled CO_2 cryogenic cooling of the machining zone. The use of CO_2 cryogenic cooling generates tool wear minimizations close to 25% of those seen using blown air or water–oil emulsion during hard turning tests. Energy is a vital production resource. For instance, around 10% of industrial energy consumption is attributable to compressed air utilization. Another element is the presence of electrical drives. Around 50% of energy usage can be ascribed to electrical drives. The remaining 40% of this industry's energy use may be attributable to heating and lighting. Kopac et al. [100] highlighted that even if output remains constant or falls, machining firms of all sizes can save money and develop their ecological performance by implementing sustainability concepts. Many industrial disasters have resulted from product overuse and manufacturing that exceeds consumer demand. As a result, warehouse inventory levels remain high. Cryogenic or high-pressure jet-assisted machining are environmentally friendly alternatives to the use of oil-based emulsions in machining. This includes power reduction, increased tool life and improved surface integrity in machining. Garetti et al. [101] highlighted that sustainability will be important to future generations. As a result, challenges of sustainability will influence all organizational sectors, including the economic, political, social and environmental. Rahim et al. [102] highlighted that sustainable machining lowers machining costs, energy consumption and waste. It would boost waste management, operational safety and employee health. Mineral-based cutting fluids are used in environmentally friendly machining. Bhanot et al. [103] investigated milling and turning sustainable machining settings. Manufacturing is competitive when economic, environmental and social variables are balanced. Metal cutting parameters are interrelated, and the parameter network for this study was developed using graph-based modularity analysis. Interactions between various components of expert systems can be parameterized. Virdi et al. [104] highlighted that manufacturing is the primary generator of economic growth, but it also has substantial environmental consequences, owing primarily to process features. A high production rate is required for the machining industry to meet demand and increase profits. To achieve production goals, feed, speed and depth of cut must all be high. High machining temperatures damage both the tool and the workpiece. Zhang [105] reviewed the MQL technology advances of using nano-enhanced biolubricant (NEBL). The advanced lubrication and heat transfer mechanisms of NEBL were revealed by quantitative comparison with MQL machining using biolubricant. Sankaranarayanan et al. [106] pointed out that because of global green machining awareness and environmental concerns, the sector is reducing its use of cutting fluid. Reduced fluid usage results in increased repair, storage, purchasing and disposal costs, as well as ecological and health concerns. While milling, non-bio-based cutting fluids pollute the environment (see Figure 2.14).

Groover et al. [107] highlighted that dry machining is an option for reducing production expenses. Here, the friction between the tool and the workpiece is substantial, resulting in an increase in temperature, which augments tool attrition and dimensional deviations. Dry cutting is only appropriate at low cutting velocities leading to a minimum output rate to prolong tool life. Virdi et al. [108] summarized that biodegradable cutting fluids must be used in place of mineral-based ones. In terms of tool attrition, power consumption and subsurface microhardness, cryogenic machining outperforms conventional machining. The heat generated by dry machining is a disadvantage. MQL reduces the need for cutting fluid, however, toxicity remains a concern.

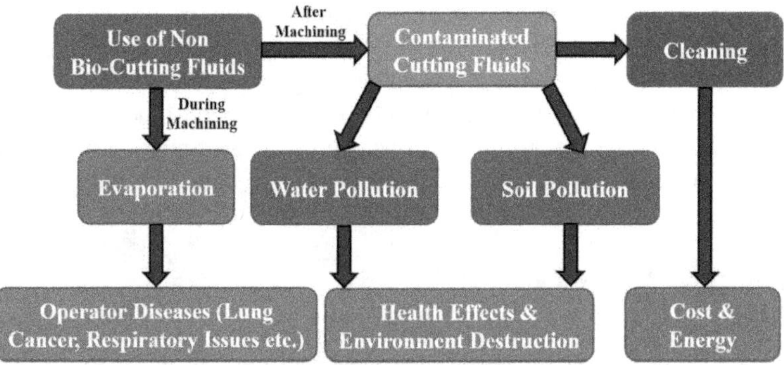

FIGURE 2.14 Environmental issues of non-bio nature cutting fluids during the machining process. (From Ref. [106].)

2.6 INFERENCES

- This chapter is expected to add to the corpus of knowledge about the effects of turning conditions on the machining distinctiveness of industrially important hard materials. It is suggested that, in the future, various levels of depth of cut be examined in order to regulate the surface roughness while turning of hardened steels.
- This chapter sheds light on the practice of sustainable machining in current manufacturing facilities. Additional future enhancements are suggested, which will aid in the development of more sustainable cooling and lubrication environments for machining processes.
- MQL is environmentally sustainable, as evidenced by extensive study on its use in different machining processes and its conjunction with other lubricating procedures.
- Further in-depth research is required to optimize cutting fluid compositions. Investigations into minimum quantity cutting fluid should also highlight minimizing power consumption, improving cooling capability so as to generate sustainable and cleaner manufacturing conditions.
- MQL, known as near-dry machining (NDM), emerges as a favourable replacement for conventional cooling methods. It is one of the most sustainable manufacturing techniques. Utilizing MQL leads to a minimization of cutting coolant, thereby minimizing its environmental impact. The adoption of NDM, being environmentally friendly, is the preferred choice for future machining.
- The economics of environmentally friendly cooling and lubrication systems must be highlighted, as they play a critical role in influencing industrial sectors to reject conventional practices in favour of environmentally friendly alternatives.
- To make intelligent and smart production for Industry 4.0, it is necessary to mix powerful computational algorithms, data science and sustainable machining and control parameters.

- Several sectors struggle to balance economic and environmental sustainability. This study recommends new green machining research directions that could have an impact on sustainable manufacturing in the future.
- It is well recognized that various cooling approaches such as nanofluids in MQL sustainable machining with various coolants and with proper environmental performance analysis lead to eco-friendly manufacturing processes.

REFERENCES

[1] S. Chinchanikar, S. S. Kore, P. Hujare, A review on nanofluids in minimum quantity lubrication machining. *Journal of Manufacturing Processes*, 68, 56–70, 2021.

[2] M. Sarikaya, M. K. Gupta, I. Tomaz, M. Danish, M. Mia, S. Rubaiee, M. Jamil, D. Y. Pimenov, N. Khanna, Cooling techniques to improve the machinability and sustainability of light-weight alloys: A state-of-the-art review. *Journal of Manufacturing Processes*, 62, 179–201, 2021.

[3] V. S. Sharma, G. Singh, K. Sorby, A review on minimum quantity lubrication for machining processes. Materials and Manufacturing Processes, 30(8), 935–953, 2014.

[4] T. He, N. Liu, H., Xia, L. Wu, Y. Zhang, Li, D., & Y. Chen, Progress and trend of minimum quantity lubrication (MQL): A comprehensive review. *Journal of Cleaner Production*, 135809, 2022.

[5] B. Boswell, M.N. Islam, I.J. Davies, Y.R. Ginting, A.K. Ong, A review identifying the effectiveness of minimum quantity lubrication (MQL) during conventional machining. *The International Journal of Advanced Manufacturing Technology*, 92, 321–340. 2017.

[6] R. S. Chandel, R. Kumar, J. Kapoor, Sustainability aspects of machining operations: A summary of concepts. *Materials Today: Proceedings*, 50, 716–727, 2022.

[7] G. W. A. Kui, S. Islam, M.M. Reddy, N. Khandoker, V. L. C. Chen, Recent progress and evolution of coolant usages in conventional machining methods: A comprehensive review. *The International Journal of Advanced Manufacturing Technology*, 119(1–2), 3–40, 2022.

[8] G. Singh, S. Singh, V. Aggarwal, Sustainable machining aspects of minimum quantity lubrication during turning of EN-31 steel utilizing singular and dual lubrication nozzles. In Sustainable Development Through Engineering Innovations: Select Proceedings of SDEI 2020. Springer, Singapore, 527–538, 2021.

[9] G. Zhu, S. Yuan, B. Chen, Numerical and experimental optimizations of nozzle distance in minimum quantity lubrication (MQL) milling process. *The International Journal of Advanced Manufacturing Technology*, 101, 565–578, 2019.

[10] G. Zhu, S. Yuan, X. Kong, C. Zhang, B. Chen, Experimental observation of oil mist penetration ability in minimum quantity lubrication (MQL) spray. *Journal of Mechanical Science and Technology*, 34, 3217–3225, 2020.

[11] H. Sohrabpoor, S. P. Khanghah, R. Teimouri, Investigation of lubricant condition and machining parameters while turning of AISI 4340. *The International Journal of Advanced Manufacturing Technology*, 76, 2099–2116, 2015.

[12] R. Fernando, J. Gamage, H. Karunathilake, Sustainable machining: Environmental performance analysis of turning. *International Journal of Sustainable Engineering*, 15(1), 15–34, 2022.

[13] O. Özbek, H. Saruhan, H, The effect of vibration and cutting zone temperature on surface roughness and tool wear in eco-friendly MQL turning of AISI D2. *Journal of Materials Research and Technology*, 9(3), 2762–2772, 2020.

[14] N. Boubekri, V. Shaikh, Minimum quantity lubrication (MQL) in machining: Benefits and drawbacks. *Journal of Industrial and Intelligent Information*, 3(3), 2015.

[15] A. T. Abbas, M. Abubakr, A. Elkaseer, M. M. El Rayes, M. L. Mohammed, H. Hegab, Towards an adaptive design of quality, productivity and economic aspects when machining AISI 4340 steel with Wiper inserts. *IEEE Access*, 8, 159206–159219, 2020.

[16] U. M. R. Paturi, N. S. Reddy, Progress of machinability on the machining of Inconel 718: A comprehensive review on the perception of cleaner machining. *Cleaner Engineering and Technology*, 5, 100323, 2021.

[17] P. Kawade, S. Bokade, Review of cooling techniques used in metal cutting processes. Advances in Materials and Processing Technologies, 1–46, 2022.

[18] R. Katna, M. Suhaib, N. Agrawal, Performance of non-edible oils as cutting fluids for green manufacturing. *Materials and Manufacturing Processes*, 1–18, 2022.

[19] R. Bag, A. Panda, A. K. Sahoo, A concise review on environmental sustainable machining conditions of hard part materials. *Materials Today: Proceedings*, 62(6), 3724–3728, 2022.

[20] G. S. Goindi, P. Sarkar, Dry machining: A step towards sustainable machining –Challenges and future directions. *Journal of Cleaner Production*, 165, 1557–1571, 2017.

[21] P. S. Sreejith, B. K. A. Ngoi, Dry machining: Machining of the future. *Journal of Materials Processing Technology*, 101(1–3), 287–291, 2000.

[22] V. Derflinger, H. Brändle, H. Zimmermann, New hard/lubricant coating for dry machining. Surface and Coatings Technology, 113, 286–92, 1999.

[23] K. R. Haapala, F. Zhao, J. Camelio, J. W. Sutherland, S. J. Skerlos, D. A. Dornfeld, I. S. Jawahir, A. F. Clarens, J. L. Rickli, A review of engineering research in sustainable manufacturing. *Journal of Manufacturing Science and Engineering*, 135(4), 2013.

[24] S. Debnath, M. M. Reddy, Q. S. Yi, Environmental friendly cutting fluids and cooling techniques in machining: A review. *Journal of Cleaner Production*, 83, 33–47, 2014.

[25] K. Mondal, S. Das, B. Mandal, D. Sarkar, An investigation on turning hardened steel using different tool inserts. *Materials and Manufacturing Processes*, 31(13), 1770–1781. 2016.

[26] L. Tang, Y. Sun, B. Li, J. Shen, G. Meng, Wear performance and mechanisms of PCBN tool in dry hard turning of AISI D2 hardened steel. *Tribology International*, 132, 228–236, 2019.

[27] A. T. Abbas, M. M. El Rayes, M. Luqman, N. Naeim, H. Hegab, A. Elkaseer, On the assessment of surface quality and productivity aspects in precision hard turning of AISI 4340 steel alloy: Relative performance of wiper vs. *conventional inserts. Materials*, 13(9), 2036, 2020.

[28] W. Jomaa, V. Songmene, P. Bocher, An investigation of machining-induced residual stresses and microstructure of induction-hardened AISI 4340 steel. *Materials and Manufacturing Processes*, 31(7), 838–844, 2016.

[29] E. Öztürk, FEM and statistical-based assessment of AISI-4140 dry hard turning using micro-textured insert. *Journal of Manufacturing Processes*, 81, 290–300, 2022.

[30] M. K. Dikshit, V. K. Pathak, R. Agrawal, K. K. Saxena, D. Buddhi, V. Malik, Experimental study on the surface roughness and optimization of cutting parameters in the hard turning using biocompatible TiAlN-coated and uncoated carbide inserts. *Surface Review and Letters*, 2023. https://doi.org/10.1142/S0218625X23400024

[31] K. Orra, S. K. Choudhury, Tribological aspects of various geometrically shaped micro-textures on cutting insert to improve tool life in hard turning process. *Journal of Manufacturing Processes*, 31, 502–513, 2018.

[32] C. Saikaew, P. Paengchit, A. Wisitsoraat, Machining performances of TiN+AlCrN coated WC and Al_2O_3+TiC inserts for turning of AISI 4140 steel under dry condition, *Journal of Manufacturing Processes,* 50, 412–420, 2020.

[33] İ. Asiltürk, M. Kuntoğlu, R. Binali, H. Akkuş, E. Salur, A comprehensive analysis of surface roughness, vibration, and acoustic emissions based on machine learning during hard turning of AISI 4140 steel. *Metals*, 13(2), 437, 2023.

[34] A. Das, S. R. Das, A. Panda, S. K. Patel, Experimental investigation into machinability of hardened AISI D6 steel using newly developed AlTiSiN coated carbide tools under sustainable finish dry hard turning. *Proceedings of the Institution of Mechanical Engineers, Part E: Journal of Process Mechanical Engineering*, 236(5), 1889–1905. 2022.

[35] K. Ishfaq, I. Anjum, C. I. Pruncu,. M. Amjad, M. S. Kumar, M. A. Maqsood, Progressing towards sustainable machining of steels: A detailed review. *Materials*, 14(18), 5162, 2021.

[36] D. Rath, S. Panda, K. Pal, Dry turning of AISI D3 steel using a mixed ceramic insert: A study. Proceedings of the Institution of Mechanical Engineers, Part C: *Journal of Mechanical Engineering Science*, 233(19–20), 6698–6712, 2019.

[37] B. K. Mawandiya, H. V. Patel, M. A. Makhesana, K. M. Patel, Machinability investigation of AISI 4340 steel with biodegradable oil-based MQL system. *Materials Today: Proceedings*, 59, 1–6, 2022.

[38] N. S.Bhadoria, G. Bartarya, On the improvement in process performance of ceramic inserts during hard turning in MQL environment. *Materials and Manufacturing Processes*, 37(3), 283–293, 2022.

[39] A. K. Vadaliya, A. B. Ghubade, P. Sharma, A. Kumar, Sustainable manufacturing related aspects in turning: A study on tool wear. *Optimization of Industrial Systems*, 391–401, 2022.

[40] M. Rafighi, M. Özdemir, S. Al Shehabi, M. T. Kaya, Sustainable hard turning of high chromium AISI D2 tool steel using CBN and ceramic inserts. *Transactions of the Indian Institute of Metals*, 74(7), 1639–1653, 2021.

[41] N. A. Özbek, O. Özbek, F. Kara, H. Saruhan, Effect of eco-friendly minimum quantity lubrication in hard machining of Vanadis 10: A high strength steel. Steel Research International, 93, 2100587, 2022.

[42] A. Das, S. K. Patel, B. B. Biswal, N. Sahoo, A. Pradhan, Performance evaluation of various cutting fluids using MQL technique in hard turning of AISI 4340 alloy steel. *Measurement*, 150, 107079, 2020.

[43] G. Singh, M. K. Gupta, H. Hegab, A. M. Khan, Q. Song, Z. Liu, M. Mia, M. Jamil, V. S. Sharma, M. Sarikaya, C. I. Pruncu, Progress for sustainability in the mist assisted cooling techniques: a critical review. *The International Journal of Advanced Manufacturing Technology*, 109, 345–376, 2020.

[44] M. Muaz, S. K. Choudhury, Experimental investigations and multi-objective optimization of MQL-assisted milling process for finishing of AISI 4340 steel. *Measurement*, 138, 557–569, 2019.

[45] G. Singh, V. Aggarwal, S. Singh, Critical review on ecological, economical and technological aspects of minimum quantity lubrication towards sustainable machining. *Journal of Cleaner Production*, 271, 122185, 2020.

[46] M. D. Selvam, N. M. Sivaram, A comparative study on the surface finish achieved during turning operation of AISI 4340 steel in flooded, near-dry and dry conditions. *Australian Journal of Mechanical Engineering*, 18(3), 457–466, 2020.

[47] S. Mazumder, K. Ghosh, B. K. Singh, S. S. Chakraborty, N. Mandal, Experimental and finite element analyses for high-speed machining of AISI 4340 steel with ZTA insert. *Journal of The Institution of Engineers (India): Series* C, 1–10, 2023.

[48] S. U. Gunjal, S. B. Sanap, Performance evaluation of molybdenum disulfide based cutting fluids under near-dry machining as an environment-friendly technique. In Techno-Societal 2020: Proceedings of the 3rd International Conference on Advanced Technologies for Societal Applications, Springer International Publishing, 675–681, 2021.

[49] J. Sharma, B. S. Sidhu, Investigation of effects of dry and near dry machining on AISI D2 steel using vegetable oil. *Journal of Cleaner Production*, 66, 619–623, 2014.

[50] Chetan, S. Ghosh, V. R. Paruchuri, Application of sustainable techniques in metal cutting for enhanced machinability: A review. *Journal of Cleaner Production*, 100, 17–34, 2015.

[51] R. R. Chakule, S. S. Chaudhari, P. S. Talmale, Evaluation of the effects of machining parameters on MQL based surface grinding process using response surface methodology. *Journal of Mechanical Science and Technology*, 31(8), 3907–3916, 2017.

[52] F. Pušavec, A. Stoić, J. Kopač, Sustainable machining process-myth or reality. *Strojarstvo*, 52(2), 197–204, 2010.

[53] A. Iqbal, K. A. Al-Ghamdi, G. Hussain, Effects of tool life criterion on sustainability of milling. *Journal of Cleaner Production*, 139, 1105–1117, 2016.

[54] K. A. Al-Ghamdi, A. Iqbal, A sustainability comparison between conventional and high-speed machining. *Journal of Cleaner Production*, 108, 192–206, 2015.

[55] A. M. Khan, M. Alkahtani, S. Sharma, M. Jamil, A. Iqbal, N. He, Sustainability-based holistic assessment and determination of optimal resource consumption for energy-efficient machining of hardened steel. *Journal of Cleaner Production*, 319, 128674, 2021.

[56] A. Şahinoğlu, Investigation of machinability properties of AISI H11 tool steel for sustainable manufacturing. *Proceedings of the Institution of Mechanical Engineers, Part E: Journal of Process Mechanical Engineering*, 236(6), 2717–2728, 2022.

[57] N. Khanna, J. Wadhwa, A. Pitroda, P. Shah, J. Schoop, M. Sarıkaya, Life cycle assessment of environmentally friendly initiatives for sustainable machining: A short review of current knowledge and a case study. *Sustainable Materials and Technologies*, e00413, 2022.

[58] M. Sarikaya, M. K. Gupta, I. Tomaz, G. M. Krolczyk, N. Khanna, S. Karabulut, C. Prakash, D. Buddhi, Resource savings by sustainability assessment and energy modelling methods in mechanical machining process: A critical review. *Journal of Cleaner Production*, 133403, 2022.

[59] N. Sihag, K. S. Sangwan, Development of a sustainability assessment index for machine tools. *Procedia CIRP*, 80, 156–161, 2019.

[60] J. W. Sutherland, J. S. Richter, M. J. Hutchins, D. Dornfeld, R. Dzombak, J. Mangold, S. Robinson, M.Z. Hauschild, A. Bonou, P. Schönsleben, F. Friemann, The role of manufacturing in affecting the social dimension of sustainability. *CIRP Annals*, 65(2), 689–712, 2016.

[61] O. Olufayo, V. Songmene, J. P. Kenné, M. Ayomoh, Modelling for cost and productivity optimisation in sustainable manufacturing: a case of dry versus wet machining of mould steels. *International Journal of Production Research*, 59(17), 5352–5371, 2021.

[62] A. D. Jayal, F. Badurdeen, Jr., O. W. Dillon, I. S. Jawahir, Sustainable manufacturing: Modeling and optimization challenges at the product, process and system levels. *CIRP Journal of Manufacturing Science and Technology*, 2(3), 144–152, 2010.

[63] A. D. Jayal, F. Badurdeen, Jr., O. W. Dillon, I. S. Jawahir, Sustainable manufacturing: Modeling and optimization challenges at the product, process and system levels. *CIRP Journal of Manufacturing Science and Technology*, 2(3), 144–152, 2010.

[64] G. Rotella, D. Umbrello, Jr., O. W. Dillon, I. S. Jawahir, Evaluation of process performance for sustainable hard machining. *Journal of Advanced Mechanical Design, Systems, and Manufacturing*, 6(6), 989–998, 2012.

[65] S. A. Khan, S. Anwar, K. Ishfaq, M. Z. Afzal, S. Ahmad, M. Saleh, Wear performance of modified inserts in hard turning of AISI D2 steel: A concept of one-step sustainable machining. *Journal of Manufacturing Processes*, 60, 457–469, 2020.

[66] A. Das, M. K. Gupta, S. R. Das, A. Panda, S. K. Patel, S. Padhan, Hard turning of AISI D6 steel with recently developed HSN2-TiAlxN and conventional TiCN coated carbide tools: Comparative machinability investigation and sustainability assessment. *Journal of the Brazilian Society of Mechanical Sciences and Engineering*, 44(4), 138, 2022.

[67] L. Celent, D. Bajić, S. Jozić, M. Mladineo, Hard milling process based on compressed cold air-cooling using vortex tube for sustainable and smart manufacturing. *Machines*, 11(2), 264, 2023.

[68] S. M. Ebrahimi, M. Hadad, A. Araee, S. H. Ebrahimi, Influence of machining conditions on tool wear and surface characteristics in hot turning of AISI630 steel. *The International Journal of Advanced Manufacturing Technology*, 114, 3515–3535, 2021.

[69] S. Pervaiz, S. Kannan, I. Deiab, H. Kishawy, Role of energy consumption, cutting tool and workpiece materials towards environmentally conscious machining: a comprehensive review. *Proceedings of the Institution of Mechanical Engineers, Part B: Journal of Engineering Manufacture*, 234(3), 335–354, 2020.

[70] E. Usluer, U. Emiroğlu, Y. F. Yapan, A. Uysal, M. Sarıkaya, N. Khanna, Investigation of the effects of hybrid nanofluid-MQL conditions in orthogonal turning and a sustainability assessment. Sustainable Materials and Technologies, 36, e00618, 2023. https://doi.org/10.1016/j.susmat.2023.e00618A

[71] P. B. Zaman, N. R. Dhar, Design and evaluation of an embedded double jet nozzle for MQL delivery intending machinability improvement in turning operation. *Journal of Manufacturing Processes*, 44, 179–196, 2019.

[72] B. Purvis, Y. Mao, D. Robinson, Three pillars of sustainability: In search of conceptual origins. *Sustainability Science*, 14, 681–695, 2019.

[73] L. Dash, S. Padhan, S. R. Das, Design optimization for analysis of surface integrity and chip morphology in hard turning. *Structural Engineering and Mechanics, An Int'l Journal*, 76(5), 561–578, 2020.

[74] K. K. Gajrani, D. Ram, M. R. Sankar, Biodegradation and hard machining performance comparison of eco-friendly cutting fluid and mineral oil using flood cooling and minimum quantity cutting fluid techniques. *Journal of Cleaner Production*, 165, 1420–1435, 2017.

[75] N. Khanna, P. Shah, M. Sarikaya, F. Pusavec, Energy consumption and ecological analysis of sustainable and conventional cutting fluid strategies in machining 15–5 PHSS. *Sustainable Materials and Technologies*, e00416, 2022.

[76] H. Javid, M. Jahanzaib, M. Jawad, M. A. Ali, M. U. Farooq, C. I. Pruncu, S. Hussain, Parametric analysis of turning HSLA steel under minimum quantity lubrication (MQL) and nanofluids-based minimum quantity lubrication (NF-MQL): a concept of one-step sustainable machining. *The International Journal of Advanced Manufacturing Technology*, 117(5–6), 1915–1934, 2021.

[77] P. Muthuswamy, An environment-friendly sustainable machining solution to reduce tool consumption and machining time in face milling using a novel wiper insert. *Materials Today Sustainability*, 22, 100400, 2023. https://doi.org/10.1016/j.mtsust.2023.100400

[78] C. Li, Y. Tang, L. Cui, P. Li, A quantitative approach to analyze carbon emissions of CNC-based machining systems. *Journal of Intelligent Manufacturing*, 26, 911–922, 2015.

[79] A. T. Abbas, M. K. Gupta, M. S. Soliman, M. Mia, H. Hegab, M. Luqman, D. Y. Pimenov, Sustainability assessment associated with surface roughness and power consumption characteristics in nanofluid MQL-assisted turning of AISI 1045 steel. *The International Journal of Advanced Manufacturing Technology*, 105, 1311–1327, 2019.

[80] H. A. Kishawy, H. Hegab, I. Deiab, A. Eltaggaz, Sustainability assessment during machining Ti-6Al-4V with Nano-additives-based minimum quantity lubrication. *Journal of Manufacturing and Materials Processing*, 3(3), 61, 2019.

[81] F. Pušavec, J. Kopač, Sustainability assessment: Cryogenic machining of Inconel 718. *Strojniški vestnik-Journal of Mechanical Engineering*, 57(9), 637–647, 2011.

[82] L. Li, T. Qu, Y. Liu, R. Y. Zhong, G. Xu, H. Sun, C. Ma, Sustainability assessment of intelligent manufacturing supported by digital twin. *IEEE Access*, 8, 174988–175008, 2020.

[83] M. S. Najiha, M. M. Rahman, A. R. Yusoff, Environmental impacts and hazards associated with metal working fluids and recent advances in the sustainable systems: A review. *Renewable and Sustainable Energy Reviews*, 60, 1008–1031, 2016.

[84] A. Das, M. K. Gupta, S. R. Das, A. Panda, S. K. Patel, S. Padhan, Hard turning of AISI D6 steel with recently developed HSN2-TiAlxN and conventional TiCN coated carbide tools: comparative machinability investigation and sustainability assessment. *Journal of the Brazilian Society of Mechanical Sciences and Engineering*, 44(4), 138, 2022.

[85] F. Pusavec, J. Kopac, Achieving and implementation of sustainability principles in machining processes. *Journal of Advances in Production Engineering and Management*, 3(4), 58–69, 2009.

[86] J. W. Sutherland, J. L. Rivera, K. L. Brown, M. Law, M. J. Hutchins, T. L. Jenkins, K. R. Haapala, Challenges for the manufacturing enterprise to achieve sustainable development. In: Manufacturing Systems and Technologies for the New Frontier: The 41st CIRP Conference on Manufacturing Systems, May 26–28, 2008, Tokyo, Japan (15–18), Springer, London.

[87] G. Rotella, D. Umbrello, Jr., O. W. Dillon, I. S. Jawahir, Evaluation of process performance for sustainable hard machining. *Journal of Advanced Mechanical Design, Systems, and Manufacturing*, 6(6), 989–998, 2012.

[88] R. Fernando, J. Gamage, H. Karunathilake, Sustainable machining: Environmental performance analysis of turning. *International Journal of Sustainable Engineering*, 15(1), 15–34, 2022.

[89] V. A. Balogun, P. T. Mativenga, Modelling of direct energy requirements in mechanical machining processes. *Journal of Cleaner Production*, 41, 179–186, 2013.

[90] M. F. Rajemi, P. T. Mativenga, A. Aramcharoen, Sustainable machining: Selection of optimum turning conditions based on minimum energy considerations. *Journal of Cleaner Production*, 18(10–11), 1059–1065, 2010.

[91] Scientific Applications International Corporation (SAIC), M. A. Curran, National Risk Management Research Laboratory (US), & Office of Research and Development, Environmental Protection Agency, United States. Life-Cycle Assessment: Principles and Practice, 2006.

[92] I. S. Jawahir, O. W. Dillon, K. E. Rouch, K. J. Joshi, A. Venkatachalam, I. H. Jaafar, Total life-cycle considerations in product design for sustainability: A framework for comprehensive evaluation. In: Proceedings of the 10th International Research/Expert Conference, Barcelona, Spain, (1) (10), 2006.

[93] J. B. Dahmus, T. G. Gutowski, An environmental analysis of machining. In: ASME International Mechanical Engineering Congress and Exposition, (47136), 643–652, 2004.

[94] H. Huang, G. Ameta, Computational energy estimation tools for machining operations during preliminary design. *International Journal of Sustainable Engineering*, 7(2), 130–143, 2014.

[95] S. Chaurasiya, G. Singh, Sustainability assessment comparison of cutting fluid for turning of titanium alloy grade II. *Process Integration and Optimization for Sustainability*, 1–7, 2023.

[96] D. Carou, J. A. Lozano, F. León-Mateos, A. Sartal, M. K. Gupta, An introduction to the use of life cycle assessment in machining. In: Corporate Governance for Climate Transition, Springer, Cham, 141–166, 2023.

[97] E. Benedicto, D. Carou, E. M. Rubio, Technical, economic and environmental review of the lubrication/cooling systems used in machining processes. *Procedia Engineering*, 184, 99–116, 2017.

[98] F. Pusavec, P. Krajnik, J. Kopac, Transitioning to sustainable production–Part I: application on machining technologies. *Journal of Cleaner Production*, 18(2), 174–184, 2010.

[99] Llanos, I. Urresti, D. Bilbatua, O. Zelaieta, Cryogenic CO_2 assisted hard turning of AISI 52100 with robust CO_2 delivery. *Journal of Manufacturing Processes*, 98, 254–264, 2023.

[100] J. Kopac, F. Pusavec, Concepts of sustainable machining processes. In: 13th International Research/Expert Conference 'Trends in the Development of Machinery and Associated Technology' TMT, 2009.

[101] M. Garetti, M. Taisch, Sustainable manufacturing: trends and research challenges. *Production Planning & Control*, 23(2–3), 83–104, 2012.

[102] E. A. Rahim, M. R. Ibrahim, A. A. Rahim, S. Aziz, Z. Mohid, Experimental investigation of minimum quantity lubrication (MQL) as a sustainable cooling technique. *Procedia CIRP*, 26, 351–354, 2015.

[103] N. Bhanot, P. V. Rao, S. G. Deshmukh, Sustainable manufacturing: an interaction analysis for machining parameters using graph theory. *Procedia-Social and Behavioral Sciences*, 189, 57–63, 2015.

[104] R. L. Virdi, A. Pal, S. S. Chatha, H. S. Sidhu, A review on minimum quantity lubrication technique application and challenges in grinding process using environment-friendly nanofluids. *Journal of the Brazilian Society of Mechanical Sciences and Engineering*, 45(5), 238, 2023.

[105] Y. Zhang, H. N. Li, C. Li, C. Huang, H. M. Ali, X. Xu, … Z. Said, Nano-enhanced biolubricant in sustainable manufacturing: from processability to mechanisms. *Friction*, 10(6), 803–841, 2022.

[106] R. Sankaranarayanan, N. R. J. Hynes, J. S. Kumar, J. A. J. Sujana, Random decision forest based sustainable green machining using Citrullus lanatus extract as bio-cutting fluid. Journal of Manufacturing Processes 68, 1814–1823, 2021.

[107] M.P. Groover, *Fundamentals of Modern Manufacturing*. John Wiley & Sons, Hoboken, NJ, 2002.

[108] R. L. Virdi, A. Pal, S. S. Chatha, H. S. Sidhu, A review on minimum quantity lubrication technique application and challenges in grinding process using environment-friendly nanofluids. *Journal of the Brazilian Society of Mechanical Sciences and Engineering*, 45(5), 238, 2023.

3 Cryogenic Assisted Hard Machining

3.1 INTRODUCTION

In hard turning, an immense amount of heat is generated, thus reducing the tool life and deteriorating the surface quality and surface integrity. Therefore, to augment machining performance, the cutting heat should be reduced by using suitable and sustainable cutting fluid technology. It is well known that the primary tasks of cutting fluids are lubricating the cutting interfaces and dispersing heat from the cutting interfaces. Usually, water based and oil based cutting fluids through flood cooling methodologies are utilized to fulfil these functions in the course of machining, but industries are enforced to utilize less harmful cutting fluids to meet techno-environmental standards [1]. Therefore, to overcome this problem, an alternative solution is developed, called minimum quantity lubrication (MQL). This technology works advantageously over dry and flood cooling due to its better lubricious capability, but it is not effectual in escaping the high amount of heat from the cutting zone [2]. Therefore, nowadays, to improve the cooling efficacy in machining, a sustainable cryogenic cooling technology is popularly implemented in the hard turning process.

Cryogenics technology requires scientists to produce and study extremely low temperature environments. The term cryogenics evolved from the Greek word "kryos", meaning cold. The first cryogenic laboratory was established by Heike Kamerlingh Onnes at the University of Leiden, Netherlands in 1882 [3]. Later, in 1919, Reitz [4] implemented liquefied gases (CO_2) as a cutting coolant for machining processes. Cryogenic machining was then first implemented in metal machining by CryoTech Company, USA in 1966. Subsequently, in 1968, Japanese scientists Uehara and Kumagai introduced cryogenic machining by replacing conventional lubricant with liquid gases like nitrogen, carbon dioxide and helium [5]. Consequently, the American Society of Heating, Refrigerating and Air-Conditioning Engineers (ASHRAE) formulated the guidelines for using cryogenic coolant in many applications in 1976 [6]. Cryogenic technology involves the study of the performance of materials at extremely low temperatures. According to the literature, temperatures below –153°C are said to be cryogenic [1, 3, 7]. Further, referring to a National Institute of Standards and Technology report, temperatures lower than –180°C are known as cryogenic temperatures [8]. The direct implementation of a cryogen coolant in a cutting zone during machining is known as cryogenic machining. To maximize their benefits,

DOI: 10.1201/9781003352389-3

precise penetration of cryogenic coolants into cutting interfaces is required [9]. Commonly, liquid nitrogen (LN_2) and liquid carbon dioxide (LCO_2) cryogenic fluids are implemented in machining activities. The primary advantage of using these cryogenic fluids is their propensity to evaporate under ambient conditions by taking heat from the cutting zone. Also, these fluids are environmentally beneficial as they do not produce any hazardous smoke due to their inertness characteristics.

Nowadays, cryogenic machining is a developing sustainable machining technique, having great potential to improve the efficacy of machinability of hard-to-cut metal alloys. Cryogenic machining is said to be the most environmentally favourable clean technology to reduce machining temperature and enable the achievement of superior tool life and surface finish [9, 10]. Recent research over the past decade has ensured the dominance of cryogenic cooling over dry, near-dry and wet/flood cooling in terms of better tool life, surface integrity and cutting temperature [1]. Also, cryogenic machining produces a contamination-free machined surface and reduces white layer formation; as a result, fatigue life of the machined component is enhanced [11].

3.1.1 Cryogenic Cooling Systems and Methods

Cryogenic cooling is a technology to deliver the cryogen during machining using different methods. Referring to the many researchers' recommendations [12–15], there are five ways to apply cryogen-cooling in cutting processes: (a) cryogen-pre-cooling of the workpiece, (b) indirect cryogen-cooling, (c) cryogen-jet/spraying cooling, (d) cryogen supplied at the chip breaker and (e) cryogen-treatment of the cutting tool. A schematic action plan for some of these techniques is displayed in Figure 3.1.

The prime objective of the cryogen-pre-cooling of a workpiece (Figure 3.1 (a)) is to lessen the workpiece temperature so that the ductile chip is made brittle, and thus broken chips are produced. According to metal cutting theory, chip breaking improves the machinability performance of any metal alloys [12, 14]. The indirect cryogen-cooling method is also termed cryogen tool back cooling. The objective of this cooling technology is to cool the machining zone through heat conduction from the LN_2 chamber positioned on the tool surface or tool holder (Figure 3.1 (b)). This technology of cooling is able to improve the cutting performance as only the insert is cooled. The LN_2 does not interact with the workpiece and cutting interfaces, thus preventing any possible alteration in workpiece characteristics. This allows stable cooling and improves machinability [16, 17]. The cryogen-spraying technology involves cooling the chip–tool interface by using LN_2 through nozzles (Figure 3.1 (c)). In this method, controlling the cryogen is difficult as it spreads in the wide space around the cutting zone. Thus, the consumption of cryogen per unit time is higher, which increases the machining cost [15]. Cryogen supplies for chip-breaker systems are being developed for chip breakability. In Figure 3.1 (d), the cryogen LN_2 is supplied to the surface of the produced chip to enhance the breakability of the chip. The nozzle shape, position and dimensions therefore play a crucial role, and these must be selected appropriately to cover the chip arc for cooling optimization [12]. In the cryo-treatment process, the heat-treated workpiece was placed in a nitrogen

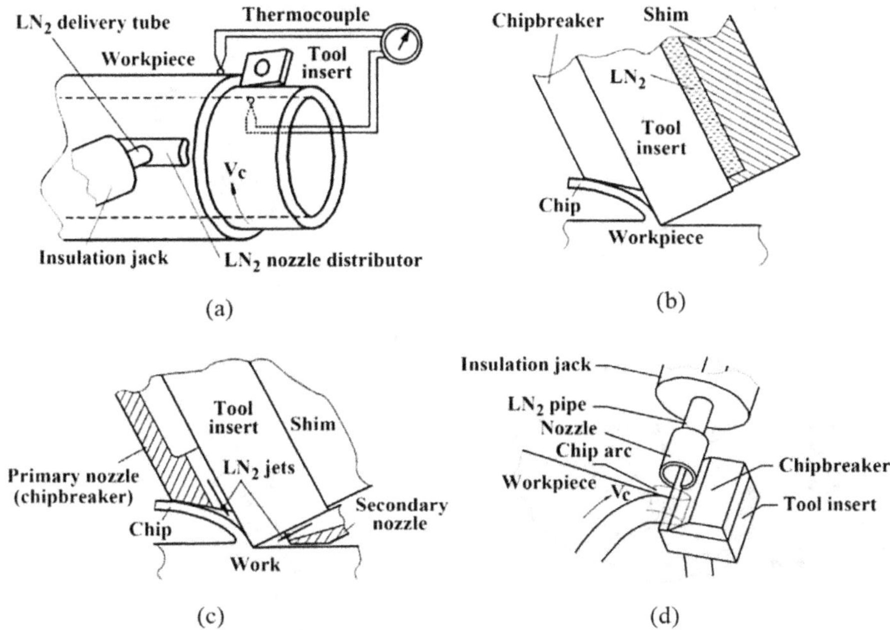

FIGURE 3.1 Cryogen-cooling methods in metal machining (a) cryogen-pre-cooling of the workpiece, (b) indirect cryogen-cooling, (c) cryogen-jet/spraying cooling and (d) cryogen supplied at the chip breaker. (From Ref. [12].)

bath and held at a constant negative temperature for a long period of time. Rafighi [18] performed cryogenic treatment of a workpiece by keeping the workpiece in a nitrogen bath at –80°C for 12 hours after heat-treatment. Further, the cryo-treated samples were heated to atmospheric temperature to achieve improved dimensional accuracy and wear resistance. Arunkarthikeyan et al. [19] applied a cryo-treatment process to a tungsten carbide cutting tool using a –196°C nitrogen bath. The holding period was varied as 24, 48 and 72 hours. The tool that was cryo-treated for 24 and 48 hours provided fine grain sized carbide particles, with improved hardness and wear resistance.

Generally, this process was commonly implemented to enhance the wear resistance and working life of materials, namely end mill and guillotine blades in manufacturing organizations. Nowadays, cryo-treated cutting tool application in the machining of hard materials is becoming popular. The scientific literature reported cryo-treated tools having improved hardness, toughness, homogeneous carbide distribution, wear resistance and working life in contrast to non-cryo-treated cutting tools [20–23]. However, the cryogenic processing system had a high investment cost, which is a drawback in terms of the applicability of this process.

The common liquids H_2, N_2, O_2, He, Ar and Ne are known as cryogenic liquids. Carbon dioxide and nitrous oxide are also categorized as cryogenic liquids. Out of these cryogenic liquids, LN_2 and LCO_2 had wide application in various cryogenic

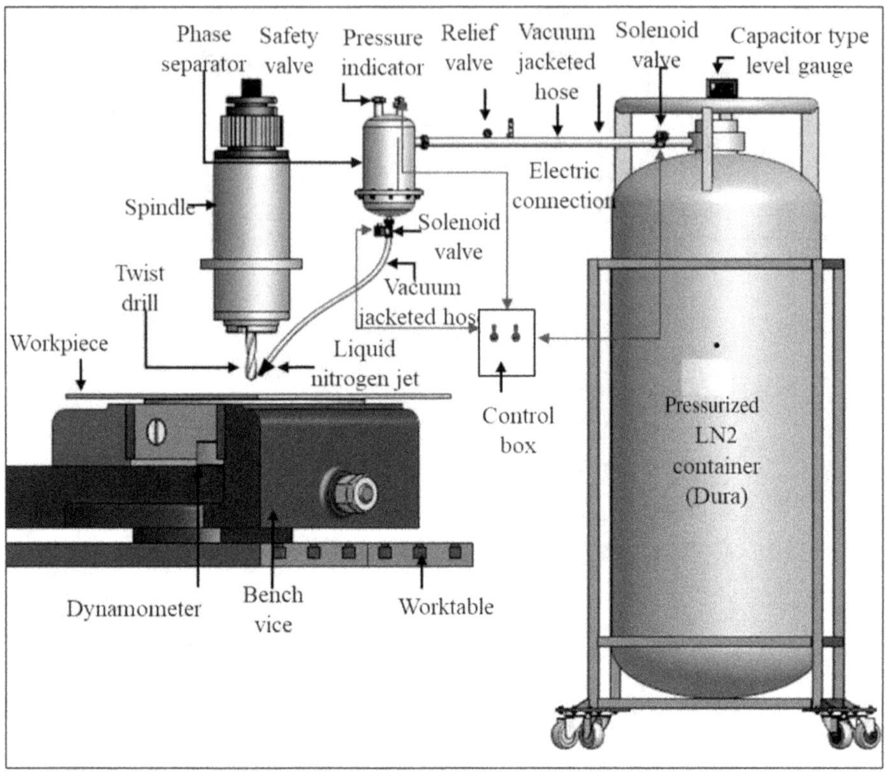

FIGURE 3.2 Schematic diagram of LN_2 cryogenic setup. (From Ref. [1].)

applications in machining activity. LN_2 is the most commonly used cryogenic coolant in machining to evaporate the evolved heat in the cutting region. LN_2 is inert, colourless, odourless, non-flammable, non-corrosive and extremely cold in nature, and therefore it is considered an eco-friendly green coolant in metal machining in comparison to conventional oil based coolants. A typical LN_2 cryogenic cooling system is disclosed in Figure 3.2 [1]. The LN_2 is stowed in an isolated tank at elevated pressure. When LN_2 is delivered into the cutting zone (atmospheric environment), it initiates boiling at –196°C and thus engrosses the heat generated from the cutting region and evaporates, removing any residuals on the machined surface and cutting tool.

Moreover, LCO_2 is an alternative cryogenic coolant to LN_2 and is utilized efficiently as a metalworking fluid in diverse machining applications. A typical CO_2 cryogenic cooling system is disclosed in Figure 3.3 [1]. As with LN_2, it is also colourless and odourless and has few toxic, pungent and acidic characteristics [24]. It has a slightly improved boiling point compared to LN_2 but its cooling effects are enough to evaporate the developed heat from the cutting region. According to Jerold and Kumar [24], LN_2 as a cryogenic coolant affects the physical characteristics of the machined sample due to its very low temperature (–196°C). Additionally, LCO_2 is much cheaper and is available in massive amounts as a waste material in many

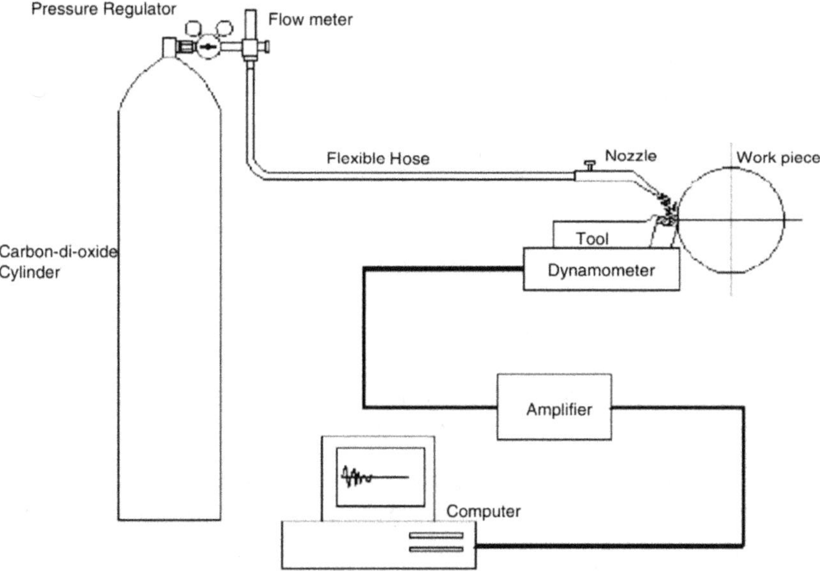

FIGURE 3.3 Schematic diagram of CO_2 cryogenic setup. (From Ref. [24].)

industries in comparison to LN_2. Therefore, in this setup (Figure 3.2), LCO_2 is stored in an insulated cylinder at moderate pressure (5.70 MPa) and atmospheric temperature. LCO_2 experiences a great drop in pressure under atmospheric conditions, resulting in cool down and magnification due to the Joule–Thomson effect. However, cutting heat is absorbed due to rapid vaporization as well as expansion of LCO_2 in the air without leaving any residuals at cutting interfaces. This phenomenon makes LCO_2 an appropriate choice for machining metal alloys [25]. Details of the relevant properties of both cryogenic coolants (LN_2 and LCO_2) are displayed in Table 3.1.

Conceptually, heat extraction is predominantly dependent on the convective heat transfer coefficient, which is proportional to the temperature difference between coolant and interface surface temperature. Higher temperature difference imparts more heat dissipation from the cutting region. The physics of cooling in cryogenic cooling systems is relatively different from other coolants. In the LN_2 system, when saturated coolant interacts with the metallic surfaces, the heat is being transferred through liquid-phase surface wetting and forced convection boiling, while in other cooling methods only forced convection works [26, 27]. Moreover, with controlled use of LN_2, the coolant volume was reduced, and this is also helpful to avoid undesired thermal distortion due to thermal expansion and thermal shrinkage [28]. Besides, factors like nozzle projection, nozzle distance, nozzle diameter, mass flow rate and inlet pressure significantly stimulate the cooling ability and cutting performance in machining. These parameters have a big role to play in reducing undesired tool wear [29, 30] and enhancing surface integrity [31] and cutting temperature in cryogenic machining. Pusavec et al. [32] examined the cooling competence of both LN_2 and LCO_2 cryogenic coolants. The cooling performance of these coolants is improved

TABLE 3.1
Relevant properties of LN_2 and LCO_2 for machining activity

Properties	LN_2	LCO_2
Boiling point @ 1 atm	–196°C [33]	–56.6°C [24]
Freezing point @1 atm	–210°C [34]	–78.5°C [12]
Density (kg/m^3)	808 @ boiling point [35]	1101 @ saturation point (–37°C) [36]
Thermal conductivity (W/mK)	0.135 [9]	0.147 [9]
Dynamic viscosity (cP) @ 0°C	0.0189 [37]	0.1726 [38]
Installation/handling of setup [33]	All components must be insulated for better performance	No insulation required
Cooling effects [33]	Low temperature formed during expansion at nozzle exit and tool-tip	Cooling of the storage container, feeding hosepipes and cutting tool

with increasing mass flow rate. The heat evacuation from the cutting zone depends on total heat transfer (sensible heat + latent heat) rate, while heat transfer of emulsion cooling depends on sensible heat as the coolant remained in a liquid phase during cooling. Besides, nozzle position significantly affects the cooling performance. An orthogonal position of the nozzle towards the tool-tip exhibited the highest cooling efficacy in machining.

3.2 MACHINING PERFORMANCE INVESTIGATIONS UNDER CRYOGENIC COOLING

LN_2 and LCO_2 cryogenic assisted machining are two different techniques used for cooling during machining operations. The performances of these cooling techniques in hard machining are addressed as follows:

3.2.1 Liquid Nitrogen Cryogenic Cooling

In LN_2 cryogenic machining, liquid nitrogen is used as the cooling agent. The liquid nitrogen is sprayed onto the workpiece or cutting tool, which rapidly cools the material, making it brittle and easier to machine. This technique is commonly used for machining hard and difficult-to-machine materials such as titanium alloys, Inconel and hardened steels. Biček et al. [39] comparatively studied the turning performance of normalized and hardened bearing steel (AISI 52100 steel) in dry, flood and LN_2 cryogenic cooling conditions. Normalized steel was turned using a coated carbide tool, while hardened steel was machined using a CBN tool. The volume of material removed vs tool wear was examined to compare the hard turning performances under dry and LN_2 cryogenic cooling conditions, as displayed in Figure 3.4. Undesired rapid growth in CBN tool wear was seen during cryo-hard-turning in contrast to cryogenic cooling due to the generation of higher tool vibration with the use of a tool with a too-negative rake and irrelevant tool edge chamfer. Moreover, with low temperature, the brittle nature of the CBN tool was enhanced and caused local fracture of the tool-tip. Therefore, with the combination of higher tool vibration, tool geometry alteration and increased brittleness, crumbling of the cutting tool in the form of chipping was produced, as displayed in Figure 3.5. This physics was noticed after the tool reached the initial wear (VB_{max} = 0.017 mm) and machine vibration was reduced. Referring to these discussions, it can be stated that the CBN tool may have a longer cutting life under cryogenic cooling in contrast to dry machining. Experimentally, longer tool life was also observed (Figure 3.4), and a 15% greater volume of material was removed in the cryogenic condition in comparison to dry cutting. In normalized steel machining, the tool removed a 1.284 × 10^5 mm^3 volume of material during flood cooling but 4.724 × 10^5 mm^3 in the cryogenic condition when wear of the tool reached

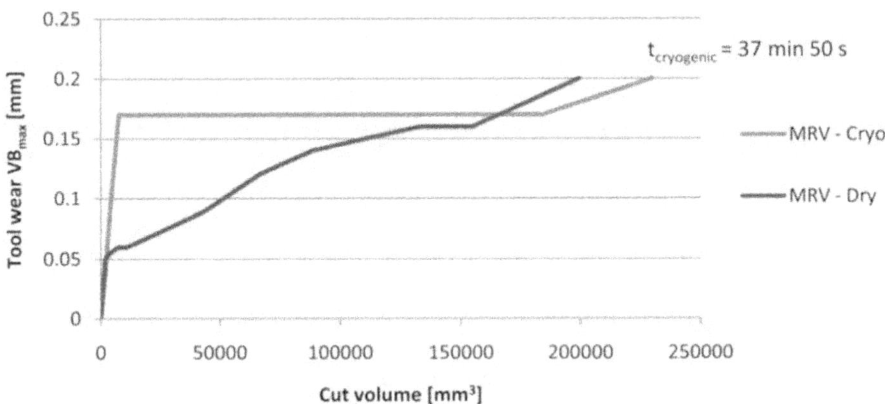

FIGURE 3.4 The progress in tool wear during machining under dry and cryogenic cooling conditions. (From Ref. [39].)

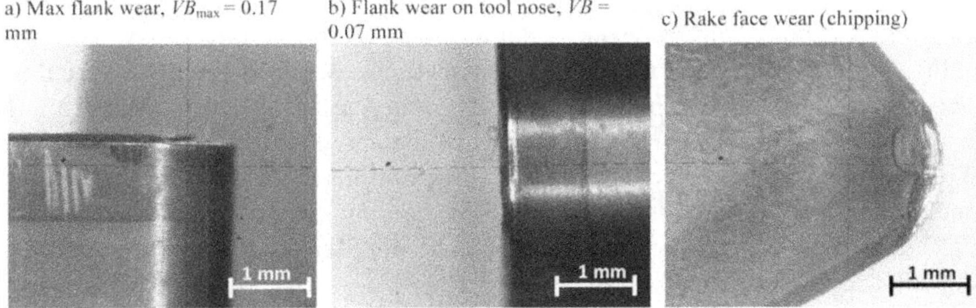

FIGURE 3.5 CBN tool wear in machining hardened steel in the cryogenic condition after machining a 120 mm length using cutting speed 100 m/min, feed 0.09 mm/rev and cutting depth 0.7 mm. (From Ref. [39].)

the maximum value (0.2 mm). Therefore, about a 268% higher volume of material was removed under cryogenic conditions.

Kumar et al. [40] compared the cryo-milling and dry milling performances of AISI 52100 grade steel at varying cutting speed (50–250 m/min) and feed (0.05–0.15 mm/tooth) and fixed depth of cut conditions. During the research, a physical vapour deposition (PVD) TiAlN coated carbide tool was implemented. The results show a 52–78% lessening in the thickness of the white layer in a cryogenic cooling (LN_2) situation due to high thermal cooling effects. A 49% drop in cutting temperature can also be observed if LN_2 cooling is used instead of dry milling. Cryogenic cooling reduces cutting forces by 28% (feed force), 28% (normal force) and 29% (axial force) in comparison to dry milling. Further, the roughness of the machined specimen is reduced by 29% in cryogenic milling as compared to dry. During the examination of the chip morphology, it was revealed that, during the cryogenic machining, silver-coloured chips of discontinuous, thin, small, serrated type were formed.

Caruso et al. [41] studied the comparative residual stress generation in machining hardened AISI 52100 steel under dry as well as cryogenic cooling conditions. They also examined the effects of cutting speed, edge geometry and microstructure alteration on residual stress. The obtained results indicated that the cryogenic cooling assisted machining induced residual profiles of less depth by limiting the white layer thickness and thus imparting a closer shift of the residual profile which enabled lower compressive residual stress in comparison to dry machining. The largest compressive residual stress was noticed just below the machined surface which minimizes the compressive area. In dry machining, further shift of residual stress was noticed, and this improves the fatigue life, but simultaneously the depth of the white layer was increased, which limits its applications.

Kumar et al. [42] compared the hard turning process on AISI 4340 grade steel (56–58 HRC) under flood (water + Servocut emulsion) and LN_2 cryogenic conditions. A TiAlN tool was used to execute the process. The comparison was based on tool life, cutting force, surface finish and power consumption results. In comparison to flood cooling, 112% higher tool life was achieved in the cryogenic condition when machining was executed at 100 m/min, while at 200 m/min speed machining, tool life under both conditions was much lower (< 3 min), as shown in Figure 3.6. Similarly, at 100 m/min, cryogenic cooling assisted machining exhibited 0.432 μm surface roughness, while it was 0.979 μm under flood cooling. However, about 55.87% lower roughness was noticed in cryogenic cooling relative to flood. Additionally, cryogenic cooling diminishes the frictions between contact surfaces, and as a result lesser cutting force was achieved in comparison to flood cooling, as displayed in Figure 3.7. In comparison to flood cooling, the percentage decrement in force was found to be 18.1, 14.03 and 11.45% at speeds of 100, 150 and 200 m/min, respectively, in the cryogenic condition. Similarly, with the use of cryogenic cooling, the power consumption was reduced in the cryogenic condition by 300, 320 and 305 W at machining speeds of 100, 150 and 200 m/min, respectively.

Huang et al. [43] focused on the machinability comparison of different test-piece hardnesses (AISI 52100) in orthogonal hard turning under dry and cryogenic cooling conditions. The hardness of the test piece was taken from 54–62 HRC while

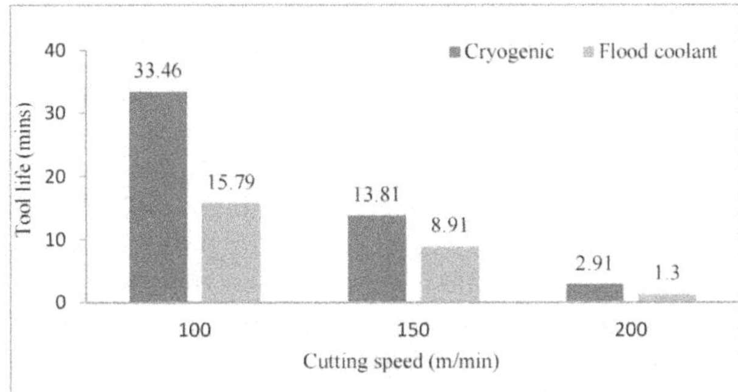

FIGURE 3.6 Obtained tool life under cryogenic and flood cooling at different cutting speeds. (From Ref. [42].)

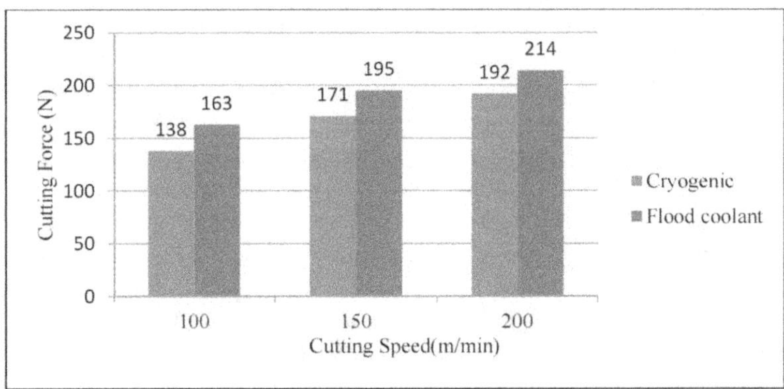

FIGURE 3.7 Obtained cutting force under cryogenic and flood cooling at different cutting speeds. (From Ref. [42].)

cutting speed changed from 75–250 m/min. The chip patterns are studied by varying speed, test-piece hardness and cutting environment. The chip pattern was found to be saw-tooth with serration. The greater serration may further cause chip segmentation. Additionally, valley and peak of chip was investigated in both cooling cutting conditions. At lower hardness (54 HRC) with smaller speed (75 m/min), the valley and peak values are similar in both conditions, while with leading speed (250 m/min) at the same hardness, the peak value was improved by 60% and the valley was lessened by 50% in comparison to dry cutting.

Khan et al. [44] evaluated the effects of emulsion cooling and LN_2 cooling approaches on sustainable parameters such as tool life, surface roughness, power consumption and production cost in external turning of AISI 52100 grade steel, and the obtained results were compared. The LN_2 approach was found to be superior over emulsion cooling for all obtained responses. The LN_2 process did not just improve the surface quality, but also reduced energy consumption by 18%, turned 66.6% greater length and created products that were 70.9% cheaper with aggressive turning inputs in comparison to emulsion. The results of this study embolden metal machining industries to adopt this type of sustainable practice in the machining shop.

Khare and Agarwal [45] executed L_9 (four factors and three levels) turning experiments on AISI 4340 steel under LN_2 cryogenic environment. The surface finishing was studied by varying tool rake angle, cutting speed, feed and cutting depth. The smallest roughness (0.097 μm) was achieved with the lowest level of input terms, while the maximum roughness (7.08 μm) was obtained at maximum speed (150 m/min), moderate feed (0.90 mm/rev), highest depth (0.6 mm) and smallest rake angle (4°). Among all input parameters, the cutting speed along with depth of cutting had the greatest impact on surface finish characteristics. In another study on turning AISI 4340 steel, Khare et al. [46] compared the turning performances under traditional coolant and LN_2 cryogenic coolant using an L_9 (four factors and three levels) experimental design. The performances were discussed using results of material removal rate (MRR) and surface roughness (Ra). In both cooling scenarios, the maximum MRR was achieved at maximum speed (250 m/min) and feed (0.135 mm/rev), moderate depth (0.4 mm) and smallest rake angle (4°). The cryo-machining exhibited far better surface quality than traditional cooling, while marginal improvement in MRR was also seen under cryogenic conditions. Leadebal Jr et al. [47] studied the surface integrity in turning AISI 52100 grade steel using a PCBN insert under dry as well as LN_2 cryogenic cooling conditions. The cutting length was taken as 25 mm in the tool feed direction. Three different nozzle position conditions were chosen for cryogenic application. In the first, the nozzle position was set towards the tool-top surface (TS). In the second, the position of the nozzle was kept towards the flank surface (FS), while in the third, both nozzle positions were used (TS/FS). In comparison to dry turning, the surface roughness was found to be smaller in cryo-machining. The lowest mean surface roughness was achieved in the TS/FS (0.20 μm) condition, succeeded by TS (0.19 μm), FS (0.18 μm) and dry (0.31 μm). Similarly, the least tool wear was corresponding to the TS/FS (0.031 mm) condition succeeded by FS (0.044 mm), TS (0.045 mm) and dry (0.055 mm). The micro-hardness of the finished piece was enhanced after cryo-cooling. The greatest improvement in micro-hardness was noticed in the TS/FS condition (28.3%) as compared to the machined micro-hardness of the job. In other cooling conditions, the micro-hardness was also increased by 17% in the FS condition, 16% in the TS condition and 7.3% in the dry condition. Therefore, as compared to cryo-machining, smaller hardness increment was seen with dry machining. In all the cooling conditions, only compressive residual stress was found. The higher residual stress was noticed in cryo-machining in comparison to dry machining.

Umbrello et al. [48] investigated the effects of cryogenic coolant on surface integrity in the hard turning of AISI 52100 steel and the results were compared with the dry turning results. The turning experiments were executed with CBN inserts, varying job hardness, different insert geometry and varying cutting speeds. The surface integrity factors, namely surface roughness, residual stress, thickness of white layer, grain size and phase transformation, were examined to explore the consequences of using cryogenic coolant and dry cutting for the surface integrity of the finished surface. The finishing quality in cryo-turning was superior to that in dry cutting, as displayed

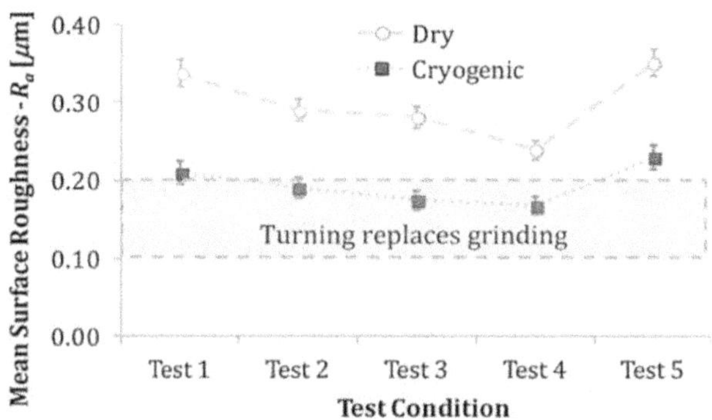

FIGURE 3.8 Surface roughness evaluation with varying turning parameters in dry and cryogenic turning. (From Ref. [48].)

in Figure 3.8. Also, a mapped region called "Turning replaces grinding" is shown in Figure 3.8, which indicates that the cryo-hard-turning can produce equivalent surface roughness to cylindrical grinding. Also, a chamfered tool produced better surface quality than a honed tool. Only compressive residual stresses in the axial direction were noticed in both cooling conditions. Relative to dry hard turning, lower residual stresses were noticed in cryo-hard-turning. Lower thickness (1 μm) of white layer was obtained in cryogenic turning in contrast to dry turning (5–8 μm). Figure 3.9 (a) and Figure 3.9 (b) graphically show the overall evaluation of surface integrity obtained in dry and cryogenic hard turning processes. Higher cutting speed along with lower hardness was more beneficial for achieving greater fatigue life and higher productivity in both cooling scenarios. Also, finer grain size was obtained in cryo-cooling in contrast to dry.

Grzesik et al. [49] reported the stimulus of cryogenic cooling on surface integrity in turning 41Cr4 hardened steel using a 60% CBN content tool. Different feed rates were implemented to study the surface characteristics of the finished part in dry as well as in cryogenic cooling. The roughness in dry hard turning was relatively lower than in cryo-hard-turning. Also, roughness was leading with feed value in both cooling conditions. Higher peaks on surface profiles due to side flow were noticed in cryo-hard-turning when the hardness and strength of the machined surface were increased. Also, finer microstructure is seen under cryogenic conditions which may cause side flow of the surface profile. Surfaces produced after cryo-hard-turning were noticeably flattened and thus showed improved bearing characteristics like negative or lower positive skewness value (–0.22 to 0.13). Under dry conditions, very sharp indentations were noticed with positive skewness value (0.26–0.32). The extreme micro-hardness in dry turning was located nearer to the cut surface, while in the cryogenic condition, it was found at a distance of 12–15 μm from the cut surface. Therefore, the micro-hardness ($HV_{0.05}$) at just below the cut surface was found to be 840 MPa, while in the cryogenic condition no white layer was produced, and thus the micro-hardness just below the cut surface was found to be 740 MPa

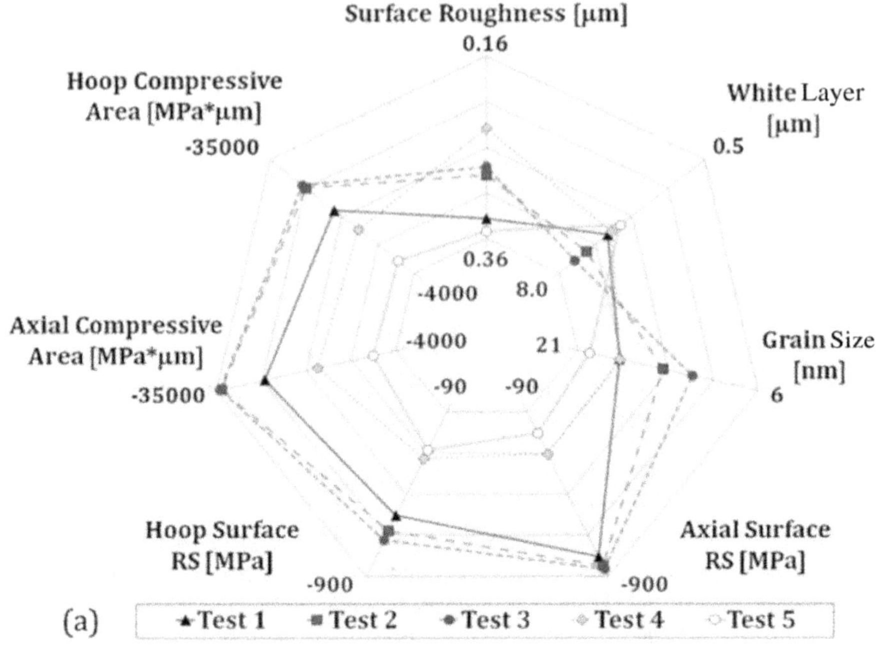

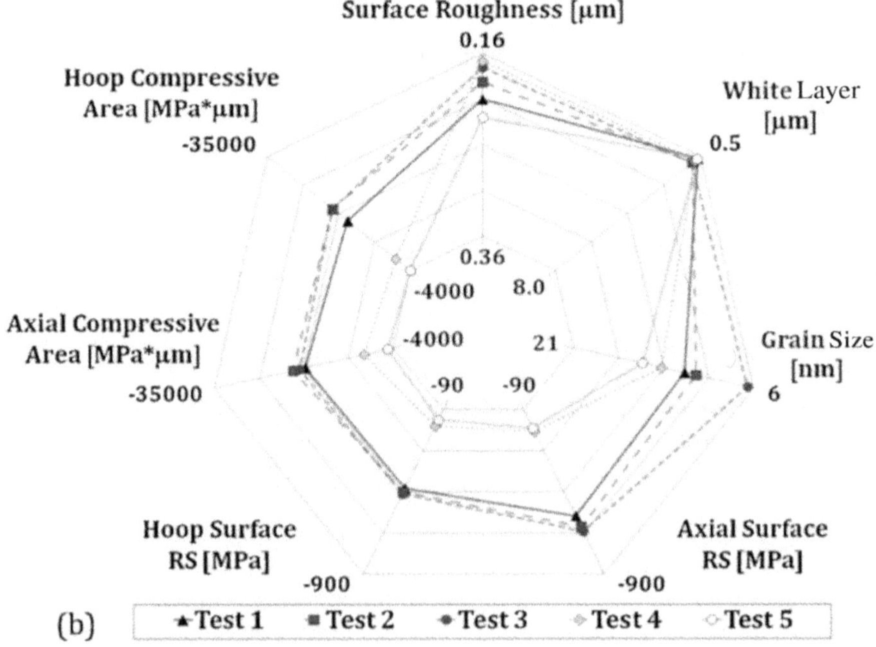

FIGURE 3.9 Overall assessment of surface integrity obtained in the (a) dry and (b) cryogenic hard turning process. (From Ref. [48].)

Ravi and Gurusamy [50] stated that cryo-machining is an ecological and advanced way to control the engendered heat in the course of machining. The end milling on heat-treated D2 steel was performed using a TiN coated tungsten carbide cutter under LN_2 cryogenic, wet and dry cutting environments. The temperature as well as cutting forces was studied in each cutting environment. The cutting heat was rising with leading cutting velocity in each cutting scenario and accelerated the insert-tip wear. The cryogenic cooling easily evaporated the developed heat, and thus temperature was reduced in contrast to in the wet and dry conditions. Similarly, the cutting force was found to be lowest in cryo-milling succeeded by wet and dry milling. In each cooling condition, the cutting force was reducing with leading cutting velocity. Gosh et al. [51] compared the tool life of a brittle tool in hard turning of A2 (62 HRC) and AISI 52100 (52–56 HRC) grade steels in dry, flood cooling and LN_2 cooling conditions. For A2 steel, the tool life of an Al_2O_3-ceramic tool with LN_2 was prolonged by more than 200% in contrast to dry as well as flood machining conditions. The prime reason for the healthier performance of the ceramic insert at elevated cutting speeds was due to potential decrease in thermal softening of the tool at leading temperatures. Additionally, with application of LN_2 coolants, as temperature in the machining zone was very low, a high temperature gradient was created between the cutting insert and the interface of the insert and the workpiece. This consequently helped efficient heat evaporation from the heat source in machining. Moreover, unexpected breakage of the tool-tip was found in flood and dry machining, which was responsible for tool failure within 4.5 min of machining in both cooling conditions. With LN_2 machining, however, only chipping had been introduced on the rake surface when the tool was stopped for further machining (13.5 min). Similarly, the LN_2 cooled alumina tool proved to have a higher tool life than the flood cooled PCBN tool in the same cutting conditions. For PCD tools, a potential decrement in cutting force was seen in comparison to other tools due to the re-sharpening and clean-shearing capability of diamond tools.

Wu et al. [52] experimentally studied the mechanics of the generation of white layers in high-speed orthogonal turning of hardened P20 grade mould steel using a PCBN insert under dry and LN_2 cryogenic cooling circumstances. Their findings revealed that the hardness of developed white layers was improved while their grain dimension was decreased in cryo-high-speed turning in comparison to dry cutting. In both cutting circumstances, there was no evidence of the material's phase modification or recrystallization alongside the white layer, but severe plastic deformation was the key reason for development of the white layer. Figure 3.10 shows the SEM micrographs of the finished surface layer after dry as well as cryo-turning with three dissimilar insert wear conditions. At wear VB = 0 mm, the workpiece surface in both cutting scenarios was a plastic deformation layer consisting of martensite deflection as well as elongation. At wear VB = 0.15 mm and 0.3 mm, the obtained surface had a fine-grained white layer structure and the parent structure of martensite had disappeared. The thickness of the developed white layer was enhanced significantly when wear of the tool-tip was increased from 0.15 to 0.3 mm.

Magadam et al. [53] investigated cutting force and tool life in cryo-turning of (45±2) HRC EN-24 steel. The experiments were performed by varying cutting speed

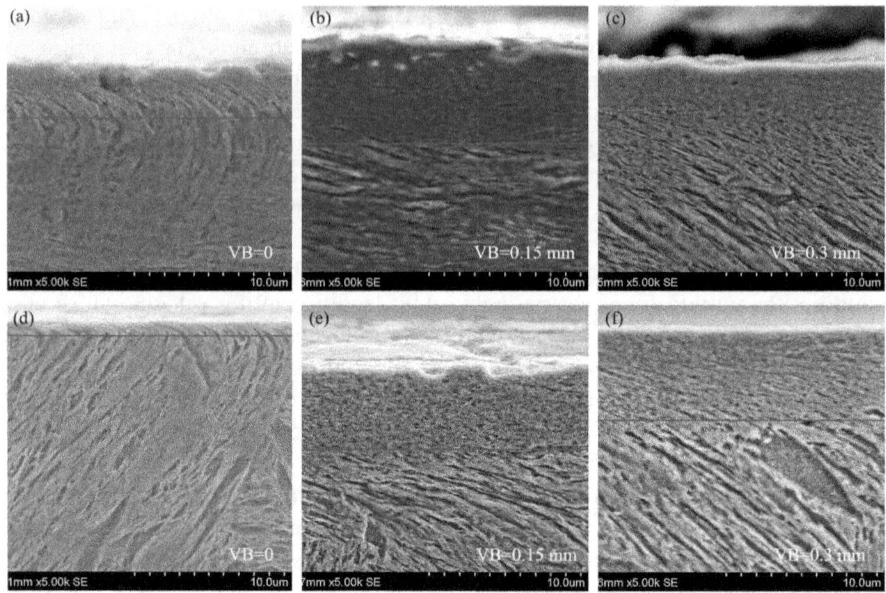

FIGURE 3.10 SEM micrographs of obtained surface layer under (a–c) dry turning at 330 m/min cutting speed and (d–f) cryo-turning at 430 m/min cutting speed. (From Ref. [52].)

(125, 160 and 200 m/min) with fixed values of feed (0.1 mm/rev) and cutting depth (0.1 mm). A TiAlNi coated carbide insert with 0.4 mm nose radius was used. The cutting force and tool life were measured under LN_2 cryogenic and flood cooling using different cutting speeds, and they were then compared. There is significant reduction in cutting force, by 15.42%, 8.22% and 7.58% at speeds of 125, 160 and 200 m/min, respectively, in contrast to flood cooling. Similarly, considerable gains in tool life, by 22.46%, 35.74% and 38.60% in cryogenic conditions at speeds of 125, 160 and 200 m/min, respectively, were found in contrast to flood cooling. Overall, a potential improvement in tool life with lessened cutting force was found in the cryogenic condition in comparison to flood cooling. Muhamad et al. [54] utilized AdvantEdge finite element method (FEM) simulation of cutting temperature in end milling of heat-treated AISI 4340 steel using a TiAlN–AlCrN insert in LN_2 cryogenic environment. The defined input parameters for this orthogonal machining simulation are cutting speed (200–300 m/min), feed (0.15–0.3 mm/tooth), axial depth (0.3–0.5 mm) and radial depth (0.2–0.5 mm). The FEM temperature during machining was found to be in the range of 576–850°C, and thus it exceeded the austenite temperature (800°C) of the hard steel used. The microstructure grained surface is changed into a nanocrystalline structured surface with grain size of 331–388 Å. The finished surface shows the presence of alpha-ferrite and martensitic phases due to rapid cooling at very low temperature. The weight% of carbon content was found to be higher on the finished surface when machining was executed at higher parameters, hence the hardness of the obtained surface was enhanced. Maximum hardness was improved by 59%, and thus it was recommended that cryo-machining had great potential to enhance the hardness of steel and thus the conventional hardening process can be eliminated. Kaynak et al. [56] had reviewed the comparative cooling performance of dry, flood, MQL and LN_2 cryogenic cooling. The report concluded that the cryo-machining induced better surface quality, improved surface topography, produced much finer grains on the surface and sub-surface of the machined component, increased the hardness of the machined component and produced a wear resistant surface with compressive residual stress, thus imparting prolonged fatigue life of the component in contrast to other cooling processes. It also facilitated a basal structure that provided an improved corrosion resistant surface. In comparison to dry and MQL machining, improved and favourable surface integrity characteristics were found with cryogenic cooling. Additionally, prolonged tool life, lower temperature, lower cutting forces and superior surface quality were found with LN_2 cryo-cooling machining.

In a recent advancement, the cryogenic coolant was applied internally through the cutting tool in the machining of hardened steel. Mia [55] studied the surface roughness, cutting force and specific cutting energy in internally cryo-cooled milling of heat-treated AISI 1060 steel using mathematical modelling and response surface multi-response optimization. Figure 3.11 shows the experimental setup, a schematic of the N_2 dewar system, the rotary applicator specifically designed to deliver the LN_2 via HSS end mill internally, a schematic view of the cutter and workpiece interaction and a schematic view of a cross-section of the end mill. The test outcomes reported that the internally cooled cryogenic cooling provided a superior performance over dry and wet cooling. The cooling method was found to be the most influential factor for surface roughness, cutting force and specific cutting energy, succeeded by feed and cutting speed. Therefore, to achieve the minimal response values, optimum machining settings of cutting speed 26 m/min and feed rate 58 mm/min and internal cryo-cooling in end milling were recommended. Islam et al. [57] performed surface milling of heat-treated EN 24 grade steel at separate speed–feed combinations resembling full factorial design (48 tests). The parametric consequences for surface roughness, tool-flank wear and cutting force were investigated through specially designed internal cryogenic cooling, dry and flood cooling. The absence of coolant in dry cutting, and the insufficient cooling and lubrication capability of the flood cooling concept, produced poorer machining results than under cryogenic coolant, and all measured quality characteristics were found to be superior in cryogenic machining. The cryogenic coolant contributed to the improved cooling and lubricant effect due to double cooling actions. Firstly, the heat evacuation rate was improved due to the primary and secondary flow of coolant within the internal slot of the tool, and secondly, the formation of a swirl flow at the existing part of the slot restricts its spreading within the work surface. The simultaneous revolution of the end mill and internally supplied coolant, centrifugal buoyancy and conduction–convection heat transfer may also be operative to enhance the cutting performance. The tool hardness was potentially enhanced due to cryogenic coolant in cutting, and thus lower wear was noticed in contrast to dry and flood milling. The cutter edge was broken in dry and flood cooling, but in cryogenic cooling only tool flaking occurred. Also, longer tool life was seen in cryo-milling in contrast to dry and wet milling, as shown in Figure 3.12. A potential decrement in cutting force was seen in cryo-machining in comparison to dry and flood cooling, as displayed in Figure 3.12. It might be possible due to a considerable

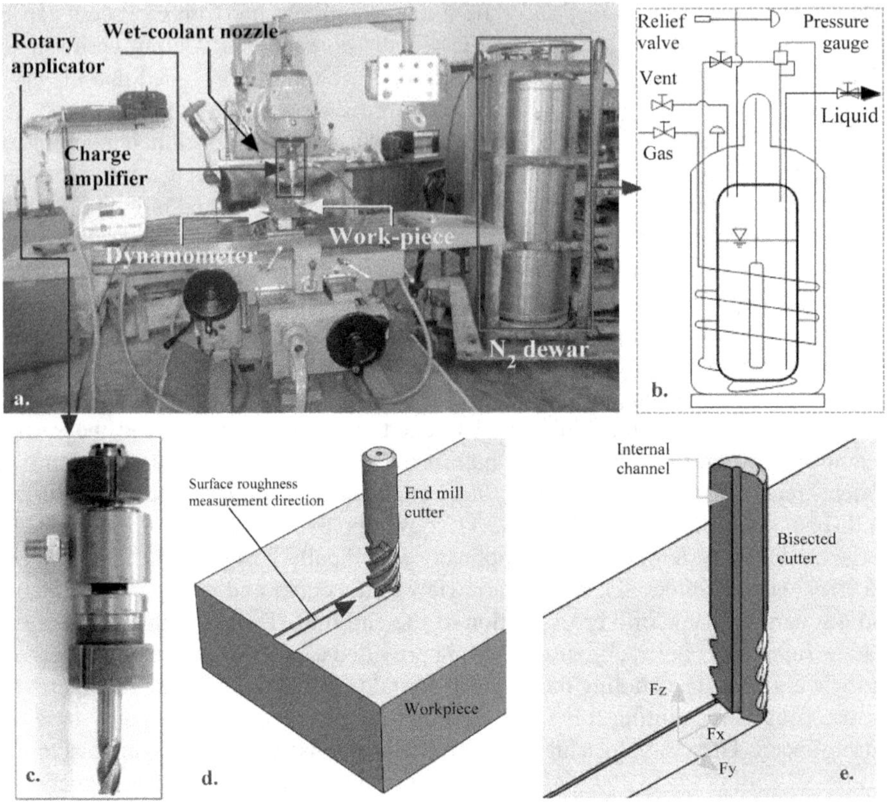

FIGURE 3.11 (a) Milling setup, (b) schematic of the N_2 dewar system, (c) specifically designed rotary applicator to deliver the LN_2 via end mill, (d) schematic view of cutter and workpiece interaction and (e) schematic view of a cross-section of the end mill. (From Ref. [55].)

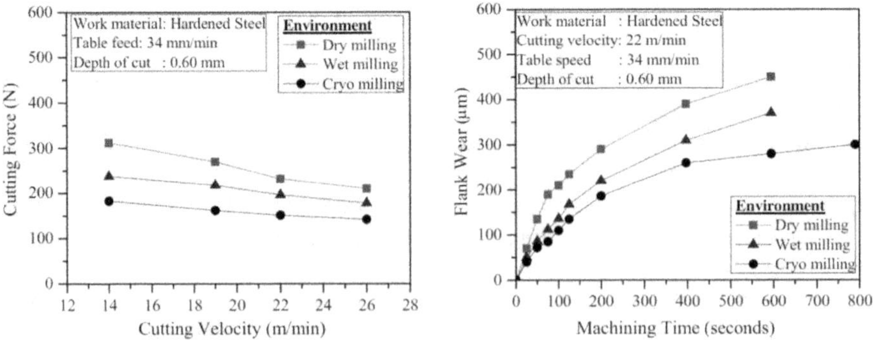

FIGURE 3.12 Comparative milling performance in dry, wet and cryogenic cooling: (a) cutting force and (b) flank wear. (From Ref. [12].)

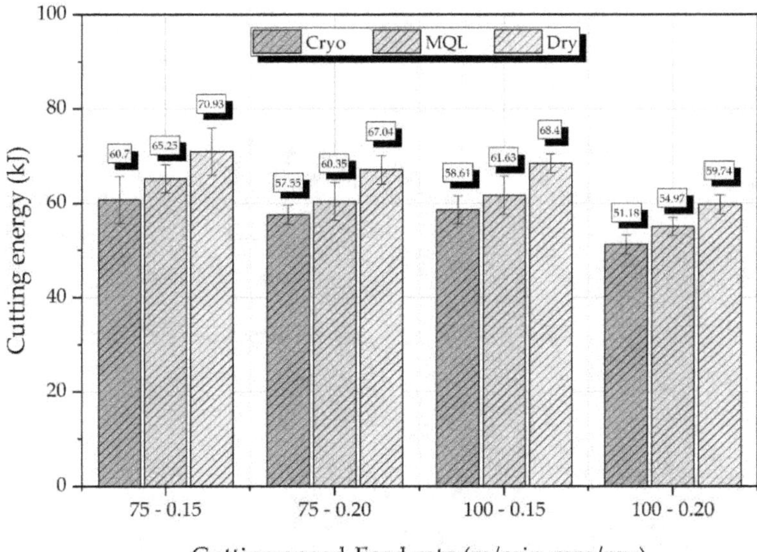

FIGURE 3.13 Comparative cutting energy results at different milling parameters. (From Ref. [58].)

reduction in tool–workpiece interface friction, reduced cutting heat, absence of built-up-edge (BUE) and effective cooling by swirl flow of the internally supplied coolant at the channel outlet [12].

In another study, Lu et al. [59] implemented LN_2 cryogenic cooling in end milling of martensitic heat resistant steel. The cutting force decreased considerably when the flow rate of LN_2 was increased. The tool life of an end mill tool was greatly enhanced (up to 20%) as compared to under conventional coolant. Furthermore, when MQL was combined with cryogenic cooling, the tool life was enhanced by 5% and the surface roughness decreased by 8% as compared to with LN_2 cooling. The cutting force under LN_2 cooling also decreased when cutting speed was improved from 100 m/min to 250 m/min. Therefore, it can be summarized that the LN_2 cooling can be more suitable for high-speed machining. The hybrid cooling (MQL+LN_2) was found to be more advantageous than LN_2 cooling as the machinability of the steel was improved.

Usca et al. [58] studied the effects of different cooling techniques (dry, MQL and LN_2 cryogenic) in gear manufacturing of AISI 5140 steel using CNC milling. The milling performance under these cooling scenarios was compared based on energy consumption, tool wear, surface roughness and chip morphology. The results showed that the cryogenic cooling exhibited improved performance over MQL and dry cutting. The lowest cutting energy (51.18 kJ) was found when milling was executed at 100 m/min speed and 0.2 mm/rev feed under cryogenic cooling conditions (Figure 3.13). Also, with leading feed rate, the cutting energy was reduced under all cooling conditions, as displayed in Figure 3.13. The flank and crater wear was seen to be lowest under cryogenic conditions in comparison to dry and MQL, as

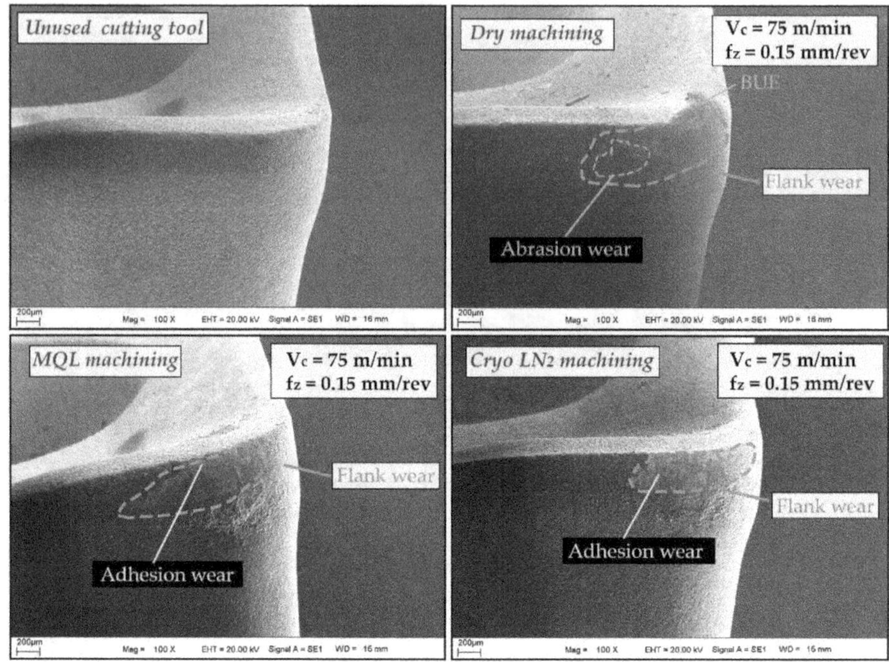

FIGURE 3.14 Tool wear comparison in dry, MQL and cryogenic cooling techniques. (From Ref. [58].)

displayed in Figure 3.14. The main wear mechanisms were traced to be abrasive and adhesive only. Further, more severe serration was seen in dry milling, as displayed in Figure 3.15. The curl radius of chips was reduced under the cryogenic condition, as shown in Figure 3.15.

3.2.2 Liquid Carbon Dioxide Cryogenic Cooling

Carbon dioxide is used as a cooling agent in CO_2 cryogenic machining. In this cooling process, CO_2 is compressed to a high pressure and then expanded through a nozzle or valve to lessen pressure and temperature. As a result, the CO_2 cools and can be applied into the interface zone of the tool and workpiece. CO_2 cryogenic machining is frequently used for machining materials that produce high temperatures during the cutting process, such as hardened steel, titanium alloys, nickel based alloys and aluminium alloys.

Urresti et al. [60] included tool flank wear and subcritical CO_2 cryogenic cooling into a Computational Finite Difference Method (CFDM) thermal model for machining AISI 52100 hardened steel. The benefits of ecologically friendly CO_2 aided machining methods are demonstrated by the decrease of tool wear with cryogenic cooling. Furthermore, the findings support the necessity to reduce temperature at the tool–workpiece contact to avoid expedited wear of PCBN tools. The

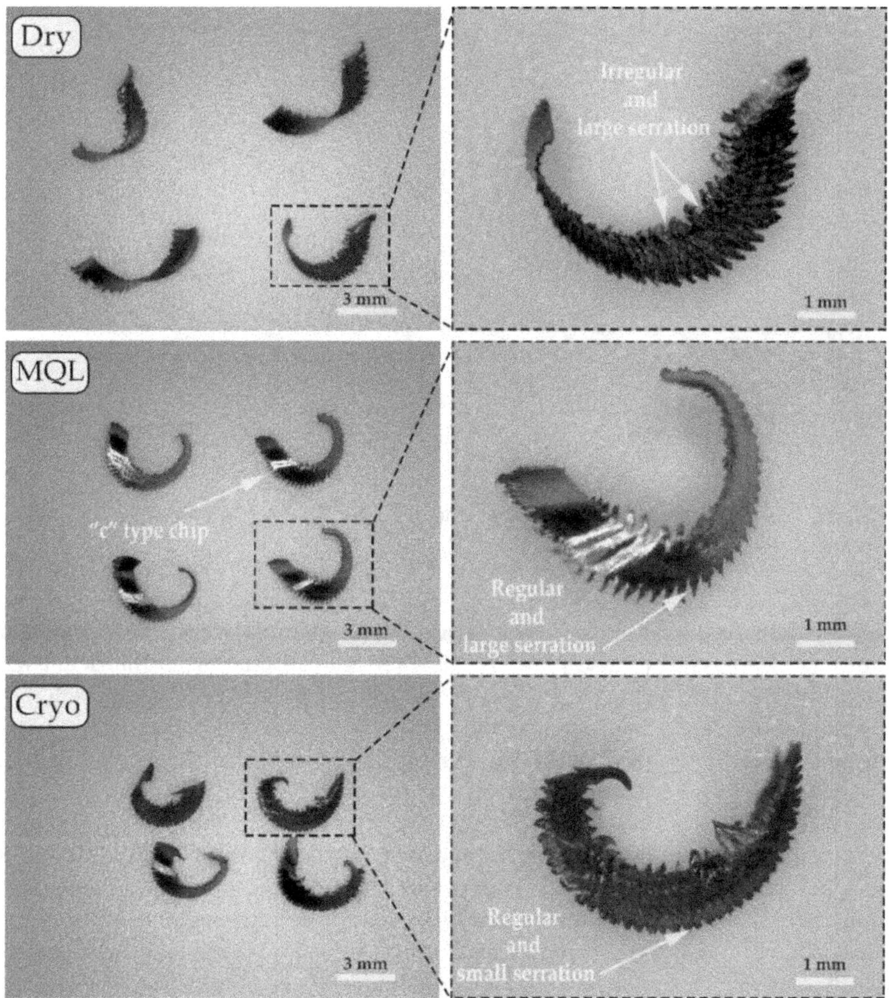

FIGURE 3.15 Comparison of chip patterns in dry, MQL and cryo-machining conditions. (From Ref. [58].)

comparison of tool-flank temperature with increasing flank wear is demonstrated in Figure 3.16 and is clearly found to be beneficial under cryo-machining. Pereira et al. [61] conducted machinability enhancement research on HSS (66 HRC) using LCO_2 cooling vs dry cutting. The responses of tool life, surface roughness and microstructure were studied. In comparison to the dry results, more than 69% improvement in tool life was noticed in the cryo-condition. Similarly, surface roughness was seen to be smaller under cryo-machining.

Kaynak and Gharibi [62] examined the performance of LCO_2, LCO_2+MQL, and LN_2 cryogenic coolants in turning AISI 4140 steel. This study discovered that

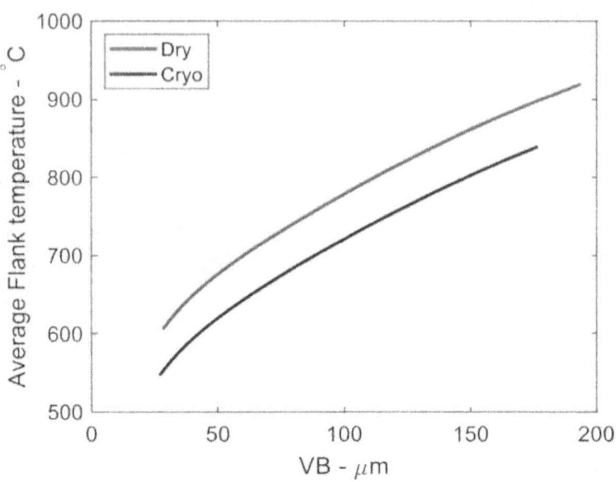

FIGURE 3.16 Comparison of tool-flank temperature under dry and LCO_2 cryo-machining. (From Ref. [60].)

LCO_2 assisted cryogenic machining created the maximum wear when compared to dry and LN_2 assisted machining. Also, when a small quantity of lubricant was given concurrently with LCO_2, it resulted in much greater tool wear, including nose and flank wear, than dry machining. It was also revealed that BUE occurs during the machining process of this material at 80 m/min cutting speed in both cryogenic cooling settings. Moreover, out of these four cutting scenarios, chip flow damage occurs only in dry machining. According to the present findings, LCO_2 assisted machining is not recommended for turning AISI 4140 steel within the specified cutting limits. Jebaraj and Kumar [63] performed end milling tests on die steel in a CO_2 environment, and the outcomes for cutting temperature, feed force, normal force, axial force, surface roughness, tool wear, chip and surface morphology, and residual stress were compared to those obtained in wet and dry circumstances. When compared to wet and dry results, the cryogenic CO_2 cooling strategy reduced temperature values by approximately 50.49% and 64.42%, respectively. Similarly, CO_2 cooling reduced all components of cutting forces when compared to dry and wet conditions. Moreover, surface roughness levels decreased by approximately 59.81% and 67.15%, respectively, in comparison to wet and dry circumstances. Gowthaman et al. [64] used an uncoated tungsten carbide cutting tool to investigate the machinability of hardened AISI 4340 steel (48 HRC) in dry, cryogenic liquid nitrogen and cryogenic liquid carbon dioxide chilling conditions. The trial results showed that cryogenic cooling effectively diminished cutting temperature, surface roughness and flank wear. The cryogenic LN_2 machined zone had the highest micro-hardness value when compared to the dry and cryogenic LCO_2 cutting conditions. Overall, when compared to dry and cryogenic LCO_2, the efficacy of the cryogenic LN_2 environment in the turning process was successful and productive.

3.2.3 Machining Performance Under Cryogenic MQL Cooling Conditions

In further advancements in cryo-technology, the cryo-minimal quantity lubrication (CMQL) technique is receiving the attention of many researchers. Nowadays, CMQL assisted machining is said to be a new type of green cutting technology which uses the advantages of cryogenic cooling and micro-lubrication cooling [65]. Liu et al. [66] stated that the CMQL cooling strategy had advantages over cryogenic cooling or MQL alone. In the CMQL system, cryogenic air and the MQL concept were utilized to get the advantages of effective cooling by cryogenic air and efficient lubrication by MQL [67]. The blending of cryogenic air with MQL enables the formation of micro-sized droplets of lubricants. When the atomizing pressure is raised, the mean diameter of droplets reduces and, as a result, the numbers of high-speed droplets are increased, the surface heat transfer coefficient is enhanced, and better cooling effects are exhibited [68, 69]. Moreover, the cryogenic air exhibited a potential decrement in MQL lubricant temperature and a significant increment in lubricant viscosity, which enables the easier adhesion of micro-droplets of lubricant to the tool-tip, thus creating a stable lubricating thin film between the tool-tip and the workpiece and reducing cutting forces, resulting in greater tool life in machining. Additionally, cryogenic air leads to more atomized lubricant micro-droplets at the surface in comparison to conventional metalworking coolants, thus as the temperature of these droplets is lower they can absorb a lot of cutting heat from the shearing zone [67].

Zhang et al. [67] implemented this CMQL cooling technology in surface milling of H13 grade steel by using three different types of internal cooling cutter (double helical slot, single straight slot and double straight slot). The tool wear and cutting force were investigated under this hybrid concept. The results revealed that milling with double straight slot cutter exhibited superior performance in extending tool life and reducing cutting forces in comparison with other types of cutters. The gradual wear and fracture of the cutter edge were found to be the dominant types of wear which affect the tool wear and cutting forces. The forces improved with escalating tool wear. In the perception of frugality and ecological safety, the double straight slot internal cooling cutter was endorsed in the cutting of H13 steel in a CMQL cooling environment. Zhang et al. [70] compared the performance of hard milling of H13 steel (40–55 HRC) under dry and CMQL cooling conditions. The tool life in CMQL and dry conditions was found to be 905 and 325 min, respectively. However, it can be said that about 178% higher tool life was achieved in CMQL in comparison to dry milling. Also, micro-chipping of the cutting edge was seen in both cooling conditions, but it was delayed under CMQL (starting after 500 min machining) as compared to dry (starting after 325 min machining). Additionally, the white layer on the finished surface was studied under both cooling scenarios. At optimal conditions with CMQL machining, the white layer almost disappeared. The results also ensured that the mechanical effect actively influenced the white layer formation rather than thermal effects. Also, the development of the white layer was closely related to flank wear as it was increasing with flank wear.

Zou et al. [65] examined the machinability behaviour of die steel using a diamond turning tool under flood, cryogenic, MQL and CMQL cooling conditions. CMQL

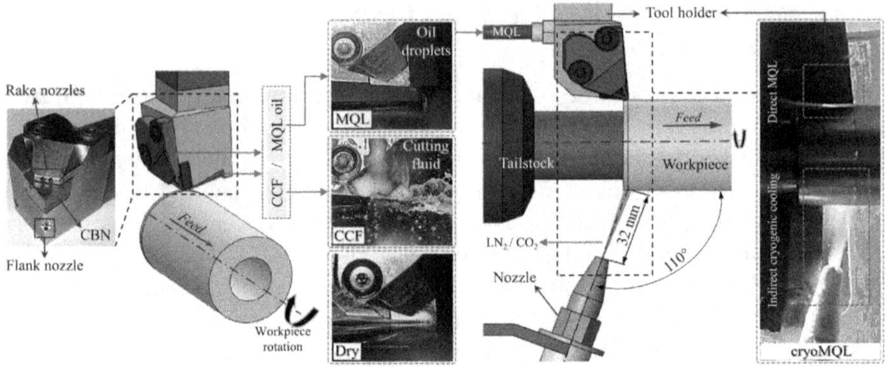

FIGURE 3.17 Schematic view of MQL, CCF, dry and cryo-MQL machining setups. (From Ref. [71].)

assisted turning provided the best performance among these cooling technologies. Under CMQL machining, the maximum temperature was recorded as the lowest among all considered cooling strategies due to the enhanced heat dissipation area meaning that cutting heat was significantly removed by forced convection while, at the same time, the atomized oil droplets vaporized by absorbing the heat formed in the cutting zone, thus providing effective cooling. Moreover, vaporized droplets successfully adsorbed on the contact zone of the workpiece and the tool create a thin lubricating film, thus reducing the frictional forces between them and significantly lowering the cutting heat. Çetindağ et al. [71] investigated the conventional and wiper CBN tool performances in hard turning of AISI 52100 bearing steel under dry, conventional cutting fluid (CCF), MQL and cryogenic (CO_2/LN_2) MQL conditions, as displayed in Figure 3.17. The turning tests were executed at a speed of 200 m/min, feed of 0.1 mm/rev and cutting depth of 0.1 mm. The performances were compared using the results for surface roughness, tool wear and residual stresses. The MQL+ CO_2 cooling machining provided the lowest wear for both types of cutting tools. Crater wear and flank wear were noticed on the tool-tip, as shown in Figure 3.18, due to abrasion and diffusion phenomena. Additionally, roughness of the part under all cooling conditions is listed in Figure 3.19, and the lowest surfaces roughness was achieved in cryo-MQL conditions. The wiper tool was found to be less than 0.12 μm, while for the conventional tool, it was greater than 0.26 μm.

3.3 SUSTAINABILITY ASSESSMENT

In recent years, industrial enterprises have recognized the importance of sustainability. The process of creating components while minimizing environmental effect, decreasing waste and encouraging social responsibility is referred to as sustainable manufacturing. It focuses on developing goods that optimize resource efficiency while minimizing negative consequences on the environment and society [72, 73]. Moreover, a sustainability evaluation must be carried out in order to evaluate

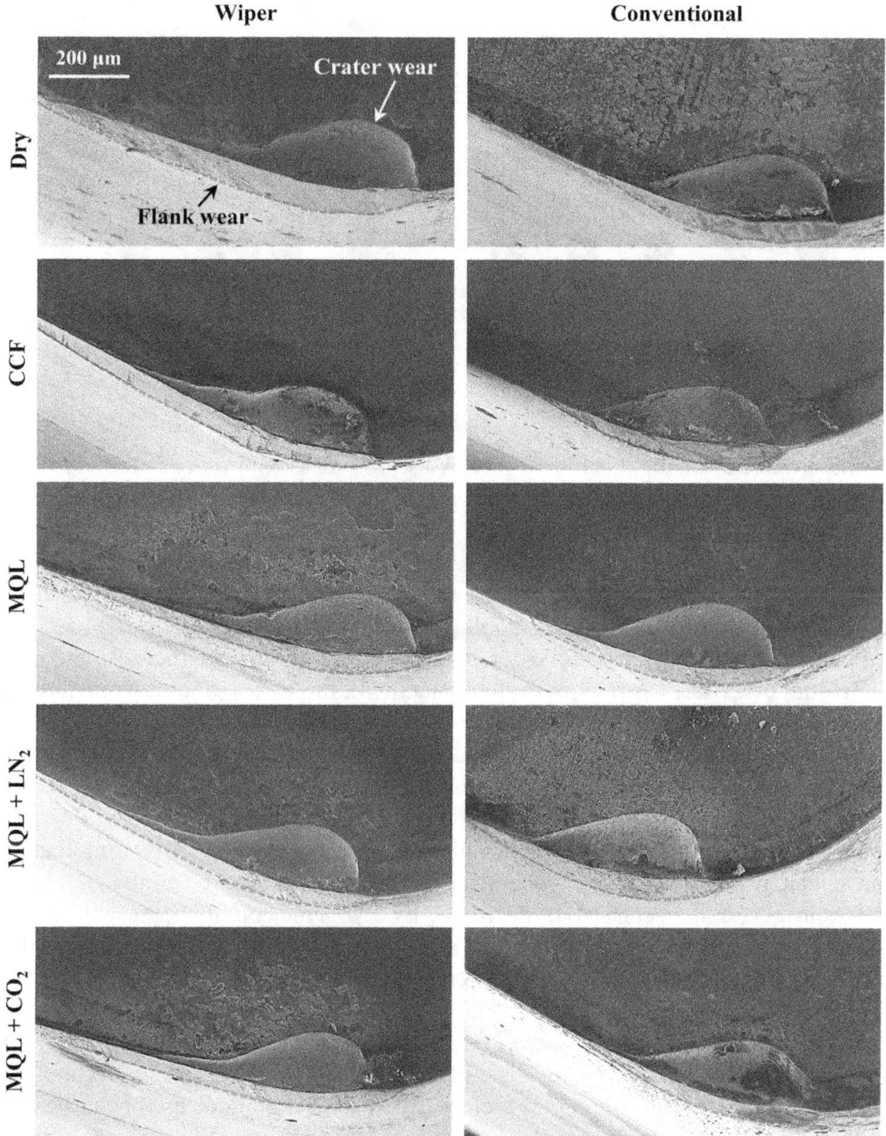

FIGURE 3.18 Crater wear and flank wear formed under different cooling techniques. (From Ref. [71].)

machining processes based on sustainability. This assessment looks at a variety of elements and components of the operation, including the environment (energy use and carbon emissions), the economics (production rate and costs), and social considerations (operator health and safety). For this assessment, the indicators must be quantitatively analyzed, and a sustainability index (SI) must be computed depending

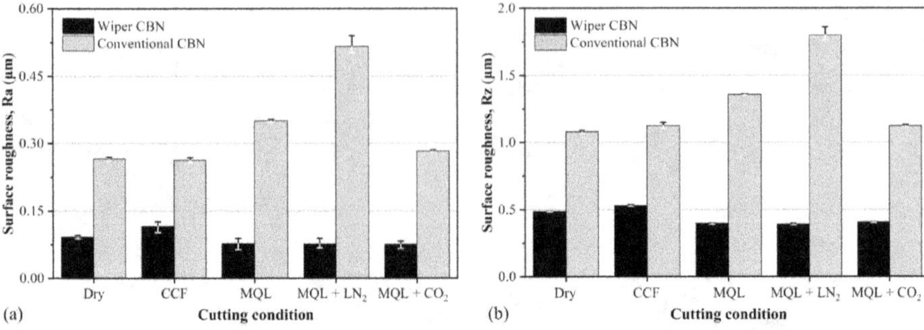

FIGURE 3.19 Comparative surface roughness obtained under different cooling conditions. (From Ref. [71].)

on the properties of the machining processes [74]. In hard turning under cryogenic cooling, sustainability evaluation entails assessing the process's influence on the environment, economy and society.

Environmental Sustainability: Cryogenic cooling may greatly reduce the quantity of coolant needed, lowering waste creation and the environmental impact of machining operations. Cryogenic cooling can also lower the cutting temperature, reducing tool wear and tear and extending tool life. Furthermore, because the cryogenic fluid evaporates as it comes into contact with the tool, it does not contaminate the workpiece or the environment, which can aid in pollution reduction [75–77].

Social Sustainability: The implementation of cryogenic cooling in machining can help to provide a safer working environment for users by lowering the amount of heat evolved during machining [75, 77]. Moreover, cryogenic cooling can lower the amount of noise generated during machining, lowering the risk of hearing damage and enhancing the operator's general health. The danger of respiratory and skin irritations in operators can also be decreased by cryogenic cooling because it does not generate any harmful fumes or mists [78, 79]. Overall, cryogenic machining offers social sustainability in terms of labour rights, operational safety, individual health and labour equity [3].

Economic Sustainability: Cryogenic cooling can bring economic benefits by lowering total machining costs. The overall cost of cryogenic machining includes machining costs, cutting fluid costs, electrical costs, cleaning costs for the finished product, recycling costs and cleaning costs for fragments and other leftover material [3]. Cryogenic cooling can lengthen the life of a cutting tool, lowering the cost of tool replacement. Cryogenic cooling can help lessen the requirement for costly coolant waste disposal as well as the expense of purchasing coolant. Additionally, because cryogenic cooling improves overall machining efficiency, it can result in higher productivity and lower manufacturing costs [80, 81].

Several studies have discussed the usage of hard machining with sustainability evaluation methods. The Pugh matrix approach combined with a Kiviat radar diagram has been used by Padhan et al. [82], Ross et al. [83], Das et al. [84] and Panda et al. [85] to investigate sustainable hard machining. They have come to the conclusion that the Pugh matrix approach in conjunction with the Kiviat radar diagram method may be used to evaluate the viability of the hard machining process's consumption of cutting inserts. In further research, the life cycle assessment (LCA) technique has been used by several writers to analyze sustainability [86–91]. LCA is a comprehensive methodology that considers all environmental exchanges across the full product development cycle, including emissions, energy, resources and industrial wastes [89]. LCA is a remarkable method for expressing the environmental impact associated with a product across its whole life cycle (from the stage of raw material acquisition through to manufacture and waste disposal) [92]. Ahmad et al. [93] reviewed the performance of three green cooling strategies in the machining of AISI 4340 steel: dry, MQL and cryogenic. In terms of surface roughness, cutting forces, and tool wear, it was discovered that cryogenic cooling outperformed dry and MQL cooling. They proposed using a hybrid cooling technology that combines cryogenic and MQL in future investigations to increase the economic sustainability of the machining process. Kim et al. [94] investigated the cost of hard turning of AISI 52100 steel with ceramic and CBN tools. The CBN tool cost 4.7 times as much as the ceramic tool. The majority of the pieces (68) were made using a cryogenic jet and the ceramic tool (flank direction). The cryogenic jet (flank direction) generated the most components (79) using the CBN tool, whereas dry cutting gave the lowest performance in terms of tool cost. According to many researchers [95, 96], cryogenic machining is more productive than conventional machining, and productivity rises steadily with increasing cutting speed. At greater cutting speeds, conventional machining productivity drastically decreases as a result of rapid tool wear. Cryogenic machining's primary advantage is achieved by reducing tool changeover times and enabling production at faster cutting speeds, which reduces the actual cutting time. Additionally, the need for cutting inserts is decreased by the extended tool life in cryogenic machining.

In conclusion, the application of cryogenic cooling in hard turning can give major environmental, social and economic benefits. Cryogenic cooling may help the overall sustainability of the manufacturing industry by lowering waste output, establishing a safer working environment and boosting machining efficiency.

3.4 INFERENCES

- The current work is a humble effort to inform readers about the various features of the cryogenic assisted hard machining process, which has become one of the most effective sustainable processes.
- Cryogen-cooling can be used in cutting operations in five different ways: (a) cryogen-pre-cooling of the workpiece, (b) indirect cryogen-cooling, (c) cryogen-jet/spraying cooling, (d) cryogen supplied at the chip breaker and (e) cryogen treatment of the cutting tool.
- Cryogenic liquids are the ubiquitous liquids H_2, N_2, O_2, He, Ar and Ne. Cryogenic liquids also include carbon dioxide and nitrous gas. Liquid nitrogen (LN_2) and carbon dioxide (LCO_2) are widely used cryogens in a variety of cryogenic machining operations.

- Cryogenic cooling applied in hard turning offers an environmentally friendly technique for manufacturing parts with high dimensional accuracy and high surface integrity.
- LN_2 was found to be the most applicable cryogen in hard turning. There have been limited studies on hard turning reporting the use of LCO_2. Therefore, detailed analysis on the performance of LCO_2 in hard machining is recommended.
- The impacts of variations in cryogen coolant flow rate and nozzle diameter need to be thoroughly explored in future research.
- Further hybridization of cryogenic machining methods with near-dry machining techniques such as MQL and nanofluid assisted MQL can improve the machinability of hardened steel.
- Although the use of LN_2 and LCO_2 in hard machining has been effective, there is still disagreement over the best cryogenic delivery methods for machining hardened steel, which restricts the technique's wide-scale adoption.
- Despite the clear advantages of cryogenic machining for the machinability of various difficult-to-machine materials, such as hardened steel, its widespread application faces particular challenges in all areas of the triple bottom line (social, economic and environmental) aspects to sustain its long-term deployment.
- To investigate the sustainable elements for industrial uses, the use of the life cycle assessment (LCA) instrument in cryo-hard machining is recommended for future research.

REFERENCES

[1] N. Khanna, C. Agrawal, D. Yu Pimenov, A. K. Singla, A. R. Machado, L. R. R. da Silva, M. K. Gupta, M. Sarikaya, G. M. Krolczyk, Review on design and development of cryogenic machining setups for heat resistant alloys and composites. *Journal of Manufacturing Processes,* 68, 398–422, 2021.

[2] B. Boswell, M. N. Islam, I. J. Davies, Y. R. Ginting, A. K. Ong, A review identifying the effectiveness of minimum quantity lubrication (MQL) during conventional machining, *International Journal of Advance Manufacturing Technology,* 92, 321–340, 2017.

[3] D. Zindani, K. Kumar, A brief review on cryogenics in machining process, *SN Applied Sciences,* 2, 1107, 2020.

[4] G. Reitz, *Die Grosse des geistlichen und ritterschaftlichen Grundbesitzes im ehemaligen Kur-Trier.* Doctoral Thesis Rheinische Friedrich Wilhelms-Universitat Bonn, Germany, 1919.

[5] K. Uehara, S. Kumagai, Chip formation, surface roughness and cutting force in cryogenic machining, *Annals of CIRP,* 17(1), 409–416, 1968.

[6] ASHRAE Handbook—*Fundamentals American Society of Heating Refrigerating and Air-Conditioning Engineers Inc.*, Atlanta Georgia, USA, 2010.

[7] C. L. Lim, N. M. Adam, K. A. Ahmad, A. Akhmedov, Cryogenic pipe flow simulation for liquid nitrogen. *Journal of Mechanical Engineering*, 2(2), 179–198, 2017.

[8] B. Gunston, *The Cambridge Aerospace Dictionary.* Cambridge University Press, New York, NY, 2004.

[9] P. Saha, R. Narain, R. Kumar, D. Singhal, A. Panda, A. K Sahoo, D. Das, Cryogenics as a cleaner cooling strategy for machining applications: A concise review, 23(8), 129–141, 2022.

[10] N. R. Dhar, S. Paul, A. B. Chattopadhyay, Role of cryogenic cooling on cutting temperature turning steel. *Journal of Manufacturing Science and Engineering, Transactions ASME*, 124(1), 146–154, 2002.

[11] Z. Zurecki, R. Ghosh, J. H. Frey, Investigation of White Layers Formed in Conventional and Cryogenic Hard Turning of Steels. Manufacturing, ASMEDC, 211–20, 2003.

[12] S. Bayraktar, Cryogenic cooling-based sustainable machining. *High Speed Machining*, 223–241, 2020.

[13] S. Pervaiz, S. Kannan, H. A. Kishawy, An extensive review of the water consumption and cutting fluid based sustainability concerns in the metal cutting sector. *Journal of Cleaner Production*, 197, 134–153, 2018.

[14] Y. Yildiz, M. Nalbant, A review of cryogenic cooling in machining processes, *International Journal of Machine Tool Manufacturing,* 48(9), 947–964, 2008.

[15] S. Y. Hong, Y. Ding, Cooling approaches and cutting temperatures in cryogenic machining of Ti-6Al-4V, *International Journal of Machine Tool Manufacturing,* 41(10), 1417–1437, 2001.

[16] Z. Y. Wang, K. P. Rajurkar, Wear of CBN tool in turning of silicon nitride with cryogenic cooling. *International Journal of Machine Tool Manufacturing,* 37(3), 319–326, 1997.

[17] Z. Y. Wang, K. P. Rajurkar, Cryogenic machining of hard-to-cut materials. *Wear,* 239(2), 168–175, 2000.

[18] M. Rafighi, Effects of shallow cryogenic treatment on surface characteristics and machinability factors in hard turning of AISI 4140 steel. Proceedings of the Institution of Mechanical Engineers, *Part E: Journal of Process Mechanical Engineering* 236(5), 2118–2130, 2022.

[19] K. Arunkarthikeyan, K. Balamurugan, P. M. V. Rao, Studies on cryogenically treated WC-Co insert at different soaking conditions. *Materials and Manufacturing Processes*, 35(5), 545–555, 2020.

[20] B. Podgornik, V. Leskovšek, J. Vižintin, Influence of deep cryogenic treatment on tribological properties of P/M high-speed steel. *Materials and Manufacturing Processes,* 24(28), 734–738, 2009.

[21] D. Thakur, B. Ramamoorthy, L. Vijayaraghavan, Influence of different post treatments on tungsten carbide–cobalt inserts. *Materials Letters,* 62(29), 4403–4406, 2008.

[22] T. V. S. Reddy, M. V. Reddy, R. Venkatram, Machinability of C45 steel with deep cryogenic treated tungsten carbide cutting tool inserts. *International Journal of Refractory Metals and Hard Materials,* 27(1), 181–185, 2009.

[23] S. Akincioğlu, H. Gökkaya, İ. Uygur, A review of cryogenic treatment on cutting tools, *The International Journal of Advanced Manufacturing Technology*, 78, 1609–1627, 2015.

[24] B. D. Jerold, M. P. Kumar, Experimental investigation of turning AISI 1045 steel using cryogenic carbon dioxide as the cutting fluid. *Journal of Manufacturing Processes,* 13(2), 113–119, 2011.

[25] D. Fernandez, A. Sanda, I. Bengoetxea, Cryogenic milling: Study of the effect of CO2 cooling on tool wear when machining Inconel 718, grade EA1N steel and gamma TiAl. *Lubricants,* 7(1), 10, 2019.

[26] C. Konishi, I. Mudawar, Review of flow boiling and critical heat flux in microgravity. *International Journal of Heat & Mass Transfer*, 80, 469–493, 2015.

[27] Y. Wang, M. Dai, K. Liu, J. Liu, L. Han, H. Liu, Research on surface heat transfer mechanism of liquid nitrogen jet cooling in cryogenic machining. *Applied Thermal Engineering*, 179, 115607, 2020.

[28] T. Lu, R. Kudaravalli, G. Georgiou, Cryogenic machining through the spindle and tool for improved machining process performance and sustainability: Pt. *I, System Design. Procedia Manufacturing*, 21, 266–272, 2018.

[29] F. Wang, Y. Wang, H. Liu, Tool wear behavior of thermal-mechanical effect for milling Ti-6Al-4V alloy in cryogenic. *The International Journal of Advanced Manufacturing Technology*, 94(5–8), 2077–2088, 2018.

[30] Y. Ayed, G. Germain, A .P. Melsio, Impact of supply conditions of liquid nitrogen on tool wear and surface integrity when machining the Ti-6Al-4V titanium alloy. *The International Journal of Advanced Manufacturing Technology,* 93(1–4), 1199–1206, 2017.

[31] Y. Kaynak, T. Lu, I. S. Jawahir, Cryogenic machining-induced surface integrity: A review and comparison with dry, MQL, and flood-cooled machining. *Machining Science and Technology*, 18(2), 149–198, 2014.

[32] F. Pusavec, D. Grguras, M. Koch, P. Krajnik, Cooling capability of liquid nitrogen and carbon dioxide in cryogenic milling. *CIRP Annals,* 68, 73–76, 2019.

[33] K. Busch, C. Hochmuth, B. Pause, A. Stoll, R. Wertheim, Investigation of cooling and lubrication strategies for machining high-temperature alloys. *Procedia CIRP*, 41, 835–840, 2016.

[34] T. Gray, The Elements: *A Visual Exploration of Every Known Atom in the Universe*. New York, Black Dog & Leventhal Publishers, 2009. ISBN 978-1-57912-814-2.

[35] C. V. Iancu, E. R. Wright, J. B. Heymann, G. J. Jensen, A comparison of liquid nitrogen and liquid helium as cryogens for electron cryotomography. *Journal of Structural Biology*. 153(3), 231–240, 2006.

[36] Wikipedia. Carbon dioxide. https://en.wikipedia.org/wiki/Carbon_dioxide

[37] I. F. Golubev, V. I. Kurin, Untersuchung der Viskosität von Gasen bei Drücken bis 4000kgf/cm² und verschiedenen Temperaturen. *Teploenergetika,* 21, 83–85, 1974.

[38] S. A. Ulybin, V. I. Makarushkin, The viscosity of carbon dioxide at temperatures of 220–1300 K and pressures up to 300 MPa. *Teploenergetika,* 65–69, 1976.

[39] M. Biček, F. Dumont, C. Courbon, F. Pušavec, J. Rech, J. Kopač, Cryogenic machining as an alternative turning process of normalized and hardened AISI 52100 bearing steel. *Journal of Materials Processing Technology*, 212(12), 2609–2618, 2012.

[40] M. S. Kumar, P. K. Shafeer, N. S. Ross, K. Ishfaq, A. A. Adediran, A. A. Akinwande, A comprehensive machinability comparison during milling of AISI 52100 steel under dry and cryogenic cutting conditions, *Proceedings of the Institution of Mechanical Engineers, Part B: Journal of Engineering Manufacture*, 237(3), 364–376, 2023.

[41] S. Caruso, J. C. Outeiro, D. Umbrello, A. C. Batista, Residual stresses in machining of AISI 52100 steel under dry and cryogenic conditions: A brief summary. *Key Engineering Materials,* 611–612, 1236–1242, 2014.

[42] S. A. Kumar, V. G. Yoganath, P. Krishna, Machinability of hardened alloy steel using cryogenic machining. Materials Today: *Proceedings*, 5, 8159–8167, 2018.

[43] A. H. Huang, Y. Kaynak, D. Umbrello, I. S. Jawahir, Cryogenic machining of hard-to-machine material, AISI 52100: A study of chip morphology and comparison with dry machining. *Advanced Materials Research,* 500, 140–145, 2012.

[44] A. M. Khan, N. He, W. Zhao, M. Jamil, H. Xia, L. Meng, M. K. Gupta, Cryogenic-LN2 and conventional emulsion assisted machining of hardened steel: Comparison from sustainability perspective. Proceedings of the Institution of Mechanical Engineers, Part B, 235(14), 2310–2322, 2020.

[45] S. K. Khare, S. Agarwal, Optimization of machining parameters in turning of AISI 4340 steel under cryogenic condition using taguchi technique. *Procedia CIRP*, 63, 610–614, 2017.

[46] S. K. Khare, K. Sharma, G. S. Phull, V. P. Pandey, S. Agarwal, Conventional and cryogenic machining: Comparison from sustainability perspective. *Materials Today: Proceedings*, 27(2), 1743–1748, 2020.

[47] W. V. Leadebal Jr., A. C. A. de. Melo, A. J. de Oliveira, N. A. Castro, Effects of cryogenic cooling on the surface integrity in hard turning of AISI D6 steel. *Journal of the Brazilian Society of Mechanical Sciences and Engineering*, 40(1), 15, 2018.

[48] D. Umbrello, F. Micari, I. S. Jawahir, The effects of cryogenic cooling on surface integrity in hard machining: A comparison with dry machining. *CIRP Annals Manufacturing Technology,* 61, 103–106, 2012.

[49] W. Grzesik, K. Zak, M. Prazmowski, B. Storch, T. Palka, Effects of cryogenic cooling on surface layer characteristics produced by hard turning. *Archives of Materials Science and Engineering,* 54(1), 5–12, 2012.

[50] S. Ravi, P. Gurusamy, Studies on the effect of cryogenic machining of AISI D2 steel, *Materials Today: Proceedings*, 37(2), 2391–2395, 2021.

[51] R. Ghosh, Z. Zurecki, H. J. Frey, Cryogenic machining with brittle tools and effects on tool life. *Proc. of ASME International Mechanical Engineering Congress & Exposition, Washington*, 37203, 201–209, 2003.

[52] S. Wu, G. Liu, W. Zhang, W. Chen, C. Wang, Formation mechanism of white layer in the high-speed cutting of hardened steel under cryogenic liquid nitrogen cooling. *Journal of Materials Processing Technology*, 302, 117469, 2022.

[53] S. Magadam, S. A. Kumar, V. G. Yoganath, C. K. Srinivasa, T. GuruMurthy, Evaluation of tool life and cutting forces in cryogenic machining of hardened steel. *Procedia Material Science,* 5, 2542–2549, 2014.

[54] S. S. Muhamad, J. A. Ghani, C. H. C. Haron, H. Yazid, Cryogenic milling and formation of nanostructured machined surface of AISI 4340, *Nanotechnology Reviews,* 9, 1104–1117, 2020.

[55] M. Mia, Multi-response optimization of end milling parameters under through-tool cryogenic cooling condition. *Measurement*, 111, 134–145, 2017.

[56] Y. Kaynak, T. Lu, I. S. Jawahir, Cryogenic machining-induced surface integrity: A review and comparison with dry, MQL, and flood cooled machining. *Machining Science Technology*, 18(2), 149–198, 2014.

[57] A. K. Islam, M. Mia, N. R. Dhar, Effects of internal cooling by cryogenic on the machinability of hardened steel. *The International Journal of Advanced Manufacturing Technology*, 90(1–4), 11–20, 2016.

[58] U. A. Usca, M. Uzun, S. Sap, K. Giasin, D. Y. Pimenov, C. Prakash, Determination of machinability metrics of AISI 5140 steel for gear manufacturing using different cooling/lubrication conditions. *Journal of Materials Research and Technology,* 21, 893–904, 2022.

[59] Jiafeng Lu, Xiaolin Deng, Jing Tang, Xiaoyun Chen, Research on liquid nitrogen cryogenic milling of 11Cr-3Co-3W martensitic heat-resistant steel, *Industrial Lubrication and Tribology*, 75/4, 457–464, 2023.

[60] I. Urresti, I. Llanos, J. Zurbitu, O. Zelaieta, Tool wear modelling of cryogenic assisted hard turning of AISI 52100 steel. *Procedia CIRP,* 102, 494–499, 2021.

[61] O. Pereira, A. Rodríguez, A. Fernández-Valdivielso, J. Barreiro, A. I. Fernández Abia, L. N. López-de-Lacalle, Cryogenic hard turning of ASP23 steel using carbon dioxide. *Procedia Engineering,* 132, 486–491, 2015.

[62] Y. Kaynak, A. Gharibi, Progressive tool wear in cryogenic machining: The effect of liquid nitrogen and carbon dioxide. *Journal of Manufacturing and Materials Processing,* 2(2), 31, 2018.

[63] M. Jebaraj, M. P. Kumar, End milling of DIN 1.2714 die steel with cryogenic CO2 cooling, *Journal of Mechanical Science and Technology,* 33(5), 2407–2416, 2019.

[64] B. Gowthaman, S. R. Boopathy, T. Kanagaraju, Effect of LN_2 and CO_2 coolants in hard turning of AISI 4340 steel using tungsten carbide tool. *Surface Topography: Metrology and Properties*, 10(1), 015032, 2022.

[65] L. Zou, Y. Huang, M. Zhou, Effect of cryogenic minimum quantity lubrication on machinability of diamond tool in ultraprecision turning 3Cr2NiMo steel. *Materials and Manufacturing Processes*, 33(9), 943–949, 2018.

[66] M. Liu, C. Li, Y. Zhang, Q. An, M. Yang, T. Gao, S. Sharma, Cryogenic minimum quantity lubrication machining: From mechanism to application. *Frontiers of Mechanical Engineering*, 16(4), 649–697, 2021.

[67] C. Zhang, S. Zhang, X. Yan, Q. Zhang, Effects of internal cooling channel structures on cutting forces and tool life in side milling of H13 steel under cryogenic minimum quantity lubrication condition. *The International Journal of Advanced Manufacturing Technology*, 83(5–8), 975–984, 2015.

[68] K. H. Park, J. Olortegui-Yume, M. C. Yoon, P. Kwon, A study on droplets and their distribution for minimum quantity lubrication (MQL). *International Journal of Machine Tools Manufacturing,* 50(9), 824–833, 2010.

[69] S. S. Sai, K. Manoj Kumar, A. Ghosh, Assessment of spray quality from an external mix nozzle and its impact on SQL grinding performance. *International Journal of Machine Tools Manufacturing,* 89, 132–141, 2015.

[70] S. Zhang, J. Li, H. Lv, *Tool wear and formation mechanism of white layer when hard milling H13 steel under different cooling/lubrication conditions. Advances in Mechanical Engineering*, 6(949308), 2014.

[71] H. A. Çetindağ, A. Çiçek, N Uçak, The effects of CryoMQL conditions on tool wear and surface integrity in hard turning of AISI 52100 bearing steel. *Journal of Manufacturing Processes*, 56, 463–473, 2020.

[72] M. A. Rosen, H. A. Kishawy, Sustainable manufacturing and design: Concepts, practices and needs. *Sustainability,* 4(2) 154–174, 2012.

[73] M. Javaid, A. Haleem, R.P. Singh, S. Khan, R. Suman, Sustainability 4.0 and its applications in the field of manufacturing. Internet of Things and Cyber-Physical Systems, 2, 82–90, 2022.

[74] M. H. Hoghoughi, M. Farahnakian, S. Elhami, Environmental, economical, and machinability based sustainability assessment in hybrid machining process employing tool textures and solid lubricant. *Sustainable Materials and Technologies*, 34, 00511, 2022.

[75] D. Fernández, A. Sandá, I. Bengoetxea, Cryogenic milling: study of the effect of CO2 cooling on tool wear when machining Inconel 718. *Grade EA1N Steel and Gamma TiAl, Lubricants*, 7(1), 10, 2019.

[76] S. S. Muhamad, J. A. Ghani, C. H. Che Haron, H. Yazid, Wear mechanism of multi-layer coated carbide cutting tool in the milling process of AISI 4340 under cryogenic environment. *Materials*, 15(2), 524, 2022.

[77] N. R. Dhar, S. Paul, A. B. Chattopadhyay, Role of cryogenic cooling on cutting temperature in turning steel. *Journal of Manufacturing Science and Engineering*, 124(1), 146, 2002.

[78] T. C. Yap, Roles of cryogenic cooling in turning of superalloys, ferrous metals, and viscoelastic polymers. *Technologies*, 7(3), 63, 2019.

[79] S. W. M. A. I. Senevirathne, M. A. R. V. Fernando, Effect of cryogenic cooling on machining performance on hard to cut metals – A literature review. *Proceeding of National Engineering Conference*, 27, 38–46, 2012.

[80] Y. Yildiz, M. Nalbant, A review of cryogenic cooling in machining processes. *International Journal of Machine Tools & Manufacture,* 48, 947–964, 2008.

[81] A. A. Khan, M. I Ahmed, Improving tool life using cryogenic cooling. *Journal of Materials Processing Technology*, 196, 149–154, 2008.

[82] S. Padhan, L. Dash, S.K. Behera, S.R. Das, Modeling and optimization of power consumption for economic analysis, energy-saving carbon footprint analysis, and sustainability assessment in finish hard turning under graphene nanoparticle–assisted minimum quantity lubrication. *Process Integration and Optimization for Sustainability*, 4, 445–463, 2020.

[83] N. S. Ross, M. Mia, S. Anwar, G. Manimaran, M. Saleh, S. Ahmad, A hybrid approach of cooling lubrication for sustainable and optimized machining of Ni-based industrial alloy. *Journal of Cleaner of Production*, 321, 128987, 2021.

[84] A. Das, M. K. Gupta, S. R. Das, A. Panda, S. K. Patel, S. Padhan, Hard turning of AISI D6 steel with recently developed HSN2-TiAlxN and conventional TiCN coated carbide tools: Comparative machinability investigation and sustainability assessment. Journal of the Brazilian Society of Mechanical Sciences, 44, 1–25, 2022.

[85] A. Panda, S. R. Das, D. Dhupal, Machinability investigation and sustainability assessment in FDHT with coated ceramic tool. *Steel Compos International Journal of Structural Integrity,* 34, 681–698, 2020.

[86] R. Fernando, J. Gamage, H. Karunathilake, Sustainable machining: Environmental performance analysis of turning. *International Journal of Sustainable Engineering*, 15, 15–34, 2022.

[87] M. Mia, M. K. Gupta, J. A. Lozano, D. Carou, D. Y. Pimenov, G. Królczyk, N. R. Dhar, Multi-objective optimization and life cycle assessment of eco-friendly cryogenic N2 assisted turning of Ti-6Al-4V. *Journal of Cleaner Production*, 210, 121–133, 2019.

[88] A. Campitelli, J. Cristóbal, J. Fischer, B. Becker, L. Schebek, Resource efficiency analysis of lubricating strategies for machining processes using life cycle assessment methodology. *Journal of Cleaner Production*, 222, 464–475, 2019.

[89] D. A. Silva, R. A. Filleti, A. L. Christoforo, E. J. Silva, A. R. Ometto, Application of Life Cycle Assessment (LCA) and Design of Experiments (DOE) to the monitoring and control of a grinding process. *Procedia CIRP,* 29, 508–513, 2015.

[90] M. K. Gupta, Q. Song, Z. Liu, C. I. Pruncu, M. Mia, G. Singh, J. A. Lozano, D. Carou, A. M. Khan, M. Jamil, D. Y. Pimenov. Machining characteristics based life cycle assessment in eco-benign turning of pure titanium alloy. *Journal of Cleaner Production*, 251, 119598, 2020.

[91] J. Shi, J. Hu, M. Ma, H. Wang, An environmental impact analysis method of machine-tool cutting units based on LCA. *Journal of Engineering, Design and Technology,* 19, 1192–1206, 2021.

[92] R. De Souza Zanuto, A. Hassui, F. Lima, D. A. Dornfeld, Environmental impacts-based milling process planning using a life cycle assessment tool. *Journal of Cleaner Production*, 206, 349–355, 2019.

[93] A. A. Ahmad, J. A. Ghani, C. H. C. Haron, Green lubrication technique for sustainable machining of AISI 4340 alloy steel. *Journal of Tribology*, 28, 1–19, 2021.

[94] D. M. Kim, H. I. Kim, H. W. Park, Tool wear, economic costs, and CO2 emissions analysis in cryogenic assisted hard-turning process of AISI 52100 steel. *Sustainable Materials and Technologies,* 30, e00349, 2021.

[95] S. Y. Hong, Economical and ecological cryogenic machining. *Journal of Manufacturing Science Technology*, 123, 331–338, 2001.

[96] F. Pusavec, J. Kopac, Achieving and implementation of sustainability principles in machining processes. *Advances in Production Engineering & Management,* 3(4), 151–160, 2009.

4 Nanofluid Assisted Hard Machining

4.1 INTRODUCTION

Hard machining is the process of turning, milling or grinding high-hardness (over 45 HRC) metallic components [1, 2]. One of the most difficult issues in hard machining is managing the heat created throughout the operation, which can cause tool wear, surface damage and reduced precision. Therefore, in recent years, many research studies have investigated various cooling strategies to overcome these issues in hard machining. Nanofluid use is one of the emerging cooling strategies for controlling the cutting temperature in hard machining. Nanofluids have shown potential in lowering the cutting temperature in hard machining, which can improve the surface finish and tool life and reduce the cutting forces during the process.

Nanofluids are a type of fluid created by suspending nanoscale particles in a base fluid like water or oil. According to ISO/TS 27687, the dimension of a nanoscale particle is denoted by its diameter and thickness or length, and that ranges in size from 1 to 100 nm [3–6]. This traditional definition of a nanoparticle is widely accepted among nanotechnology professionals. The categorization of dimensions smaller than 100 nm stems from the phenomenon that the surface area to volume ratio rapidly increases as particle size decreases. As a result of the dominance of surface effects (e.g. interfacial interactions and dominating contributions from surface energy), the material characteristics of these new nanoparticles frequently diverge dramatically and anomalously from their bulk (conventional) values. In other words, nanoparticles have distinct features that differ greatly from their bulk properties. In-depth investigations into the transport mechanisms responsible for these abnormal behaviours are still a hot issue in the scientific community [3].

The concept of "nanofluids" was initially introduced by Choi and Eastman [7] in 1995. They disseminated a modest amount of copper nanoparticles into water and found a significant enhancement in thermal conductivity in comparison to the parent coolant water. In order to improve transport phenomena and thermophysical appearance (such as viscosity, thermal conductivity and specific heat capacity) numerous combinations of nanoparticles and base fluids have subsequently been researched [8–11]. Nanoparticles, nanofibres, nanotubes, nanowires, nanorods and nanosheets are all examples of solid nano-sized materials used for synthesis of nanofluids for cutting coolant application [4]. Nanoparticles can be metallic,

DOI: 10.1201/9781003352389-4

non-metallic, metal oxides, carbide, carbonic, ceramics, a blend of two or more nanopowders (hybrid nanoparticles), or even nanoliquid droplets [4]. The most common materials utilized in the synthesis of nanofluids are single element metals (examples: copper, iron and silver), single element oxides (examples: copper oxide, aluminium oxide, silicon oxide, zirconium oxide, graphene oxide, zinc oxide, iron oxide, magnesium oxide and titanium oxide), alloys (examples: copper–zinc, iron–nickel and silver–copper), multi-element oxides (examples: zinc–iron oxide and nickel–iron oxide), metal carbides (examples: silicon carbide, zirconium carbide and boron carbide), metal nitrides (examples: silicon nitride, aluminium nitride and titanium nitride) and carbon materials like graphite, carbon nanotubes and diamond [3, 4, 12]. These nano-sized particles can be suspended in different types of fluids such as water, ethanol, ethylene glycol, synthetic oil, mineral oil, vegetable oil and refrigerants [3].

Despite the vast range of materials that have been explored in the literature for synthesizing nanofluids, the general concept of dispersing nanoparticles in a base fluid is to enhance certain material properties of the base fluid to achieve enhanced performance in a chosen application. For example, high thermal conductivity is desired in heat transfer applications (often at the expense of higher viscosity and pump penalty), while improved load-carrying capacity and non-Newtonian rheological behaviour are preferred in lubrication applications (with a concomitant ability to dissipate heat rapidly) [3]. These considerations are critical when using nanofluids as a cutting coolant in hard machining applications. To fulfil the cooling and lubricating purpose in hard machining, the selection of suitable nanoparticles, their weight or volume and the synthesis technique are extremely difficult. Other factors, such as nanofluid stability and cost, are critical for attaining long-term employment as a cutting lubricant in hard machining and being approved as a sustainable cutting coolant [3].

In machining applications, introducing nanoparticles (100 nm in diameter) into the parent cutting fluid improved its heat transfer capabilities, thus decreasing the machining temperature in the shearing area substantially [13]. The lubricious properties of nanoparticles mingled with cutting fluid are improved by: developing a thin tribo-film in between two interacting surfaces [14], ball bearing impacts [15], the polishing effect [16], and the mending effect [17]. According to the findings of Lee et al. [18], polishing along with mending of nanofluids was the key approach responsible for the improvement in the machining results. As reported by Peng et al. [19], adhesion between a pair of interacting surfaces and anti-wear capabilities of nanoparticle-mixed cutting fluids are strengthened by four unique mechanisms: a) spherical-shaped nanoparticles roll flawlessly between two interacting surfaces, thus shifting the motion of slipping friction into rolling as well as sliding frictions, (b) nanoparticles have a tendency to mix with friction surfaces and produce a thin surface-protecting layer, (c) in the course of cutting in a nanofluid environment, the impinged nanofluid gathers on the contact surfaces, resulting in a thin tribo-film that adjusts for material loss; this phenomenon has been referred to as the mending effect. Furthermore, friction reduces the lubricating roughness of the surface due to the presence of nanoparticles between the contact surfaces; this consequence is known as the polishing effect [20], and (d) evenly dispersed nanoparticles sustain the

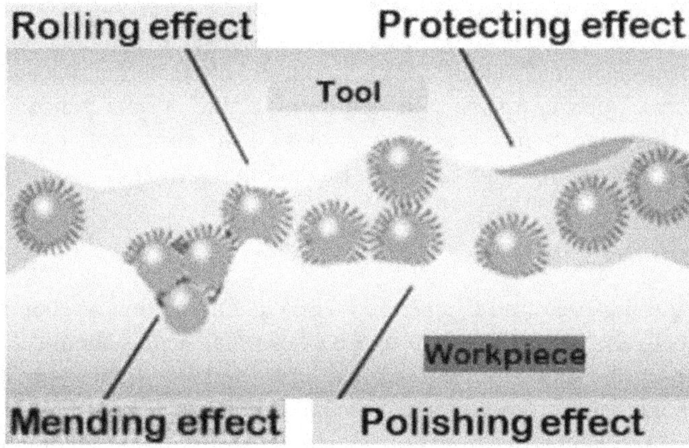

FIGURE 4.1 Mechanisms of nanoparticle interaction in a nanofluid in machining. (From Ref. [20].)

machining load (compressive) and reduce the stress concentration. These mechanisms are illustrated in Figure 4.1 [20].

Nanofluids serve as a coolant in MQL systems, and can be sprayed using MQL over the cutting zone in hard machining. The use of a nanofluid coolant can provide several advantages over traditional coolants in hard machining. The major benefits of the use of nanofluids are as follows:

- Improved heat transfer: Nanofluids have been proven to have substantially greater thermal conductivity than standard coolants, allowing them to evacuate heat from the cutting zone more effectively. This can result in lower cutting temperatures, less tool wear and a better surface quality [21].
- Reduced friction: The use of nanofluids can minimize the coefficient of friction between the tool and the workpiece, resulting in reduced cutting forces and increased tool longevity [22].
- Improved lubrication: Nanoparticles like graphite and molybdenum disulphide have been used in the synthesis of certain nanofluids to improve the lubricating characteristics of the base fluid and result in enhanced tool life [23, 24].

Overall, nanofluids have emerged as a promising cooling strategy for hard machining, with potential to improve the efficiency and effectiveness of the process.

4.2 SYNTHESIS, CHARACTERIZATION AND STABILITY OF NANO-CUTTING FLUIDS

Nano-cutting fluids are advanced lubricants used in machining operations to increase tool performance, improve surface smoothness and extend tool life. The growing need for high-precision machining techniques and sustainable production has resulted in the development of nano-cutting fluids with enhanced tribological characteristics, cooling performance and stability. The synthesis, characterization and stability of nanofluids are important aspects in determining their performance in machining operations. This chapter provides an overview of the current state of the art in the synthesis, characterization and stability of nano-cutting fluids.

4.2.1 Synthesis

Nano-cutting fluids are often synthesized using a combination of chemical, physical and biological processes. Chemical approaches entail using chemical processes to create nanoparticles and mix them with a base fluid. Physical approaches include using ultra-sonication, high-pressure homogenization and microfluidics to disperse pre-formed nanoparticles in a base fluid. Micro-organisms are used in biological ways to synthesize nanoparticles and mix them with a base fluid. The synthesis approach used is determined by parameters such as the required nano-cutting fluid qualities, the type of nanoparticles employed and the compatibility of the base fluid with the nanoparticles. The synthesis of nanofluids is a critical stage in the experimental study of nanofluids. In order to create nanofluids, various approaches have also been used. They are mostly classified into two types. The first is a one-step procedure, while the second is a two-step one.

In a one-step methodology, direct evaporation and condensation of nanoparticulate components in the base liquid are produced simultaneously to manufacture stable nanofluids. In other words, making and spreading the nanoparticles in the fluid is accomplished simultaneously. This approach reduces nanoparticle agglomeration and improves fluid stability by avoiding the drying, storing, transportation and dispersion of nanoparticles [25, 26]. The one-step approach can produce nanoparticles that are evenly disseminated and capable of remaining stably suspended in the base fluid. However, the presence of leftover reactants as a result of an incomplete reaction has been an inherent disadvantage of the technique [3].

Several one-step techniques for nanofluid preparation have been developed and are shown in Figure 4.2. Laser ablation is a quick and efficient one-step method for vaporizing metal solids and creating nanofluids. This process involves directing a

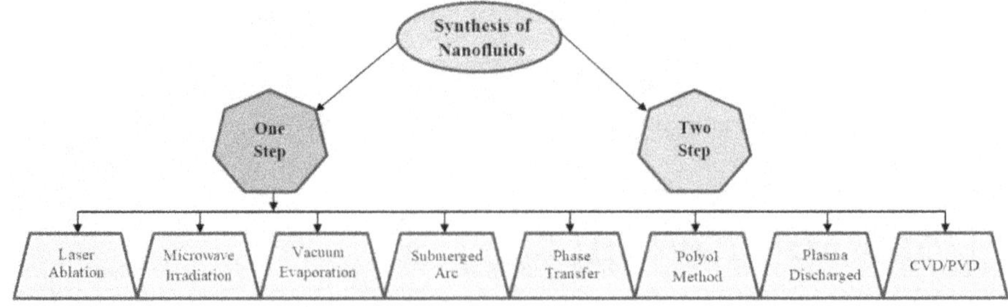

FIGURE 4.2 Nanofluid synthesis procedures.

powerful laser beam at a solid surface that remains submerged for the proper amount of time, causing the solid to melt and vaporize over the ablation threshold. A thin liquid layer close to the solid surface will likewise vaporize along with the metal in the meantime. The metal is fractured into nanoscale droplets by expanding the volume of liquid and converted to vapour, which is then super-cooled by the surrounding liquid and turned into nanoparticles [27]. This technique was utilized to synthesize a number of nanofluids containing various nanoparticles, such as Al, Cu, Sn, Si/SiC, CuO, Al_2O_3 and C particles [28–35]. For the synthesis of nanofluids in water or other organic liquids under ambient circumstances, laser ablation in the liquid is a relatively straightforward and "green" (eco-friendly) procedure compared to other approaches. For a more in-depth analysis, it is recommended to follow the work reported by Zeng [36].

Several researchers have used microwave irradiation to encourage the chemical events that lead to the production of nanoparticles in liquids [37–40]. By uniformly heating chemicals and solvents, microwave irradiation can offer a quick synthesis method that helps to hasten the chemical reaction of metal precursors as well as the formation of nanoparticles in solutions. Such characteristics produce mono-dispersed nanostructures, which is advantageous for the creation of nanofluids. In comparison to other chemical approaches, these techniques were proven to be quick and effective for creating nanofluids in one-step synthesis processes [3]. This technique was utilized to synthesize Cu [37, 38], SnO_2 [39] and ZnO [40] nanofluids.

Vacuum evaporation onto a running oil substrate (VEROS) is another single-step method to produce nanofluids. Akoh [41] created ferromagnetic metal oxide (Fe_3O_4, CoO and NiO) based nanofluids using a vacuum evaporation method. The approach produces particles in oil that have an average diameter of 2.5 nm. Wagner [42] also implemented the same technique by employing magnetron sputtering to create silver and iron nanofluids in oil. By resistively heating the source metal into cooled liquid, Eastman [43] created Cu/ethylene glycol nanofluid (10 nm Cu nanoparticles were formed in ethylene glycol with 0.5% volume concentration) and noticed negligible agglomeration.

Submerged arcing is another single-step method to synthesize nanofluids. For the first time, Tsung [44] created copper nanoparticle suspensions in deionized water using an arc-submerged method. By altering the ambient pressure throughout the synthesis, Cu nanoparticles having either coarse or fine bamboo leaf architectures (200 nm) were produced. Lo et al. [45] utilized vacuum based submerged arc technology to synthesize CuO, Cu_2O and Cu based nanofluids with different dielectric liquids. For this process, a sufficient power source is needed to produce an electric arc between 6000–12,000°C for melting and vaporizing a metal rod in the region where an arc is generated. Further, in order to create nanofluids, the vaporized metal is condensed and then dissolved in deionized water.

Phase transfer is another single-step approach to creating nanofluids. This process has been utilized for making homogeneous and stable graphene oxide colloids. After modification with the use of oleylamine, graphene oxide nanosheets were effectively transported from water to n-octane [3–4]. Yu et al. [46] also implemented this approach to create stable kerosene based Fe_3O_4 nanofluids. Oleic acid is effectively grafted onto the surface of Fe_3O_4 nanoparticles through chemisorption, allowing Fe_3O_4 nanoparticles to be kerosene compatible. The previously reported "time dependence of the thermal conductivity characteristic" does not exist in the Fe_3O_4 nanofluids generated by the phase-transfer technique.

Another one-step approach for creating nanofluids that is popular among researchers is the polyol method [47–49]. Alcohols are often diols such as ethylene glycol, propanol or diethylene glycol that are used as solvents in this process. They have the ability to dissolve a variety of inorganic precursors and also function as reducing agents. As a result, at a certain temperature, a quick nucleation and slow development of the particle is achieved. Using this technique, it has been feasible to produce metal nanoparticles and metal oxides that have a tightly controlled shape and minimal size dispersion. Although in recent years other variations have been published, such as the polyol approach using microwaves, this technique is still well known to produce metal nanoparticles and metal oxides with low size distribution and controlled shape [50, 51]. A. Guzman [52] used a microwave to create Cu nanoparticles using the polyol approach. Polyvinylpyrrolidone was added after ascorbic acid had been dissolved in ethylene glycol. The mixture was then agitated and microwave-heated at a steady rate of heat with the addition of anhydrous copper acetate ($CuAc_2$) in ethylene glycol. Cu particles in the solution are centrifuged to produce a reddish powder.

The plasma discharge method is another one-step nanofluid preparation methodology. Hsin et al. [53] produced carbon nanotubes (CNTs) by carbon plasma discharging in water. In this technique, two graphite electrodes are brought together (head to head), resulting in plasma discharge and the formation of carbon soots in aqueous solution. After plasma discharge, the water-soluble carbon soot was filtered with filter paper, cleaned with ethanol, and left to dry overnight in a 100°C oven. However, a low yield of carbon nanotubes/nanoparticles was discovered in this manner.

Nanoparticles have been created by the direct condensation of the target metal vapour in contact with a flowing liquid, and this process is known as physical vapour deposition (PVD). Controlling the vapour release and liquid flow rates allows for the creation of different particle concentrations and diameters. These techniques were developed and modified to produce mono-dispersed nanoparticles in a liquid environment. They were originally derived from the gas evaporation approach used to prepare tiny metal particles in an inert gas atmosphere [54]. To synthesize the nanofluid, a low vapour pressure base fluid and high power for metal vaporization are necessary for the standard PVD chamber. The creation of nanofluids may be done more quickly and conveniently by using a localized high temperature/heat flux approach. One typical method used to produce metal vapour with localized high energy input is the "exploding wire approach" (also referred to as "pulsed wire explosion" or "pulsed wire evaporation approach") [55]. Lee et al. [56] employed the pulsed wire evaporation approach for preparation of ethylene glycol–ZnO nanofluid. Chemical vapour deposition (CVD) was used by Akhavan [57] to produce multi-walled carbon nanotube (MWCNT) nanofluids using deionized water. Sharmin et al. [58] used CVD manufactured CNT nanoparticles to synthesize deionized base nanofluids.

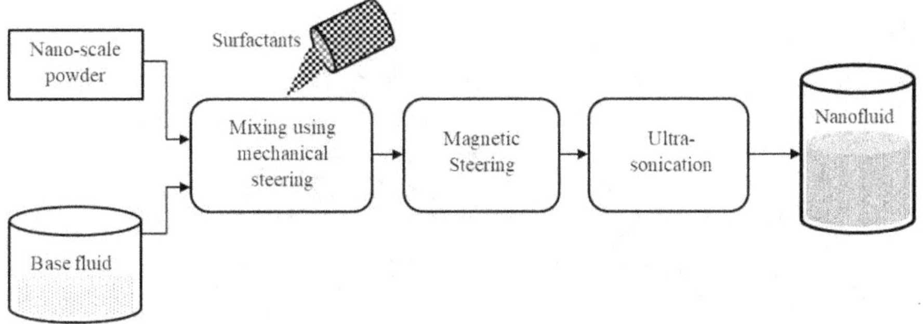

FIGURE 4.3 Two-step methodology to synthesize nanofluids.

Two-step techniques are the most cost-effective approaches for producing nanofluids on a wide scale. The majority of hard machining researchers choose this type of approach for preparing nanofluids. In a two-step procedure, the nanoparticles with suitable surfactant are dispersed into a base fluid and subjected to high-speed mechanical steering followed by magnetic steering and ultra-sonication, as displayed in Figure 4.3. This manufacturing technique is both economical and suitable for wide-scale production. The biggest disadvantage of the two-step technique is nanoparticle agglomeration and sedimentation of nanoparticles into the base fluid. Therefore, a surfactant is utilized to enhance the stability of the nanofluid. This is a highly usable commercial method for producing nanofluids for various industrial applications. Sharmin et al. [58] implemented two-step techniques to synthesize four different concentrations (0.5, 0.4, 0.3 and 0.2 % by volume) of CNT dispersed deionized water based nanofluids for machining hardened steel (42CrMo4). Das et al. [59] also implemented a two-step method to synthesize three different rice bran oil based nano-cutting fluids using copper oxide, alumina and iron oxide (Fe_2O_3) nanoparticles for hard turning applications. Yıldırım [60] utilized a two-step method to synthesize Fuchs Plantocut 10 SR based graphene nanofluids (0.5 % volume concentration) in turning hardened 420 steel. Junankar et al. [61] implemented a two-step preparation methodology to produce Cu based nanofluid in machining bearing steel.

4.2.2 Characterization

The characterization of nano-cutting fluids is greatly stimulated by nanoparticle structure (shape, size, particle interaction, coagulation, etc.). Transmission electron microscopy (TEM) and scanning electron microscopy (SEM) are effective tools for determining the size, shape and distribution of nanoparticles [62]. Das et al. [63] conducted SEM and TEM tests of four different types (copper oxide, zinc oxide, iron oxide and alumina) of nanoparticles to prepare nano-cutting fluid for hard turning of AISI 4340 steel. The SEM and TEM images for these nanoparticles are shown in Figure 4.4 and Figure 4.5, respectively. Moreover, as the sample used for characterization is often dried out, the true activities of nanoparticles in the base fluids

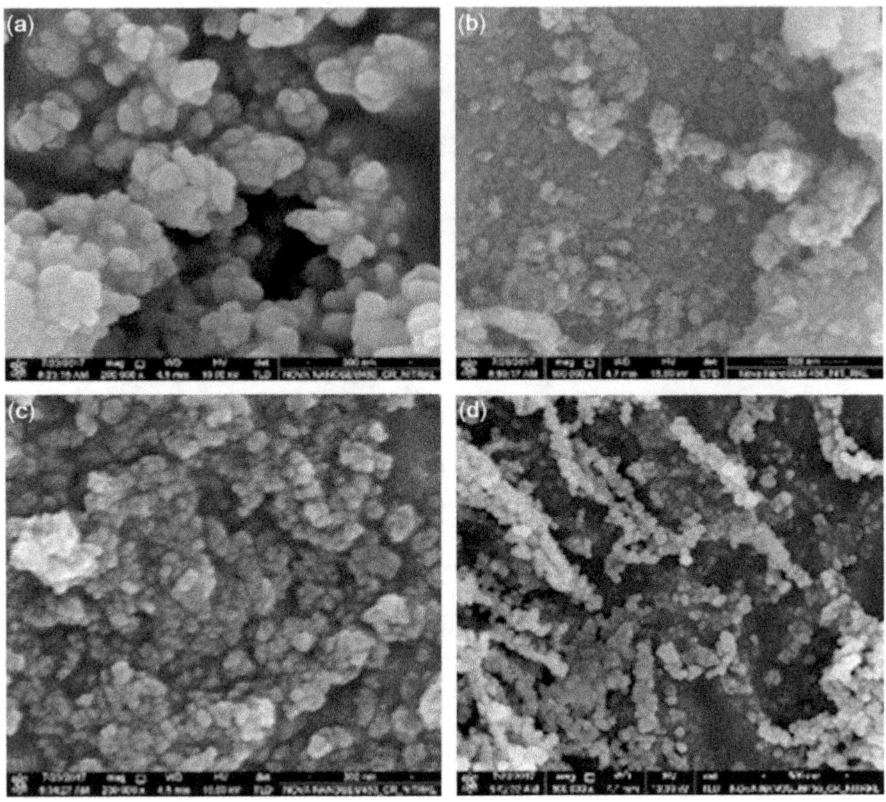

FIGURE 4.4 Nanoparticle characterization using SEM: (a) CuO, (b) ZnO, (c) Fe_2O_3 and (d) Al_2O_3. (From Ref. [63].)

are not observable. However, if the microstructure of nanofluids does not change as a result of cryogenic operations, cryo-SEM and cryo-TEM could be able to overcome this issue [62, 64]. These methods are highly useful for directly observing the nanoparticle aggregation structure in base fluids. Laser scattering is another method that uses the light intensity distribution concept to analyze nanoparticles' sizes and their distribution in a nanofluid. Another helpful technique for determining particle size, shape and distribution is SAXS (small angle X-ray scattering) [65], which looks at the elastic behaviour of X-rays when they are dispersed at short angles (typically 0.1–10°). A technique used to analyze the structure of nanoparticles in base fluids that is very similar to SAXS is small angle neutron scattering [62]. It is possible to characterize particle motion and dynamic structure using additional scattering techniques such as X-ray photon correlation spectroscopy [62], laser correlation spectroscopy [64], and others.

There are a few important basic properties of any cutting fluid that affect its characteristic behaviour. The four important properties of thermal conductivity, viscosity, density and wettability are presented here to examine the characteristics of

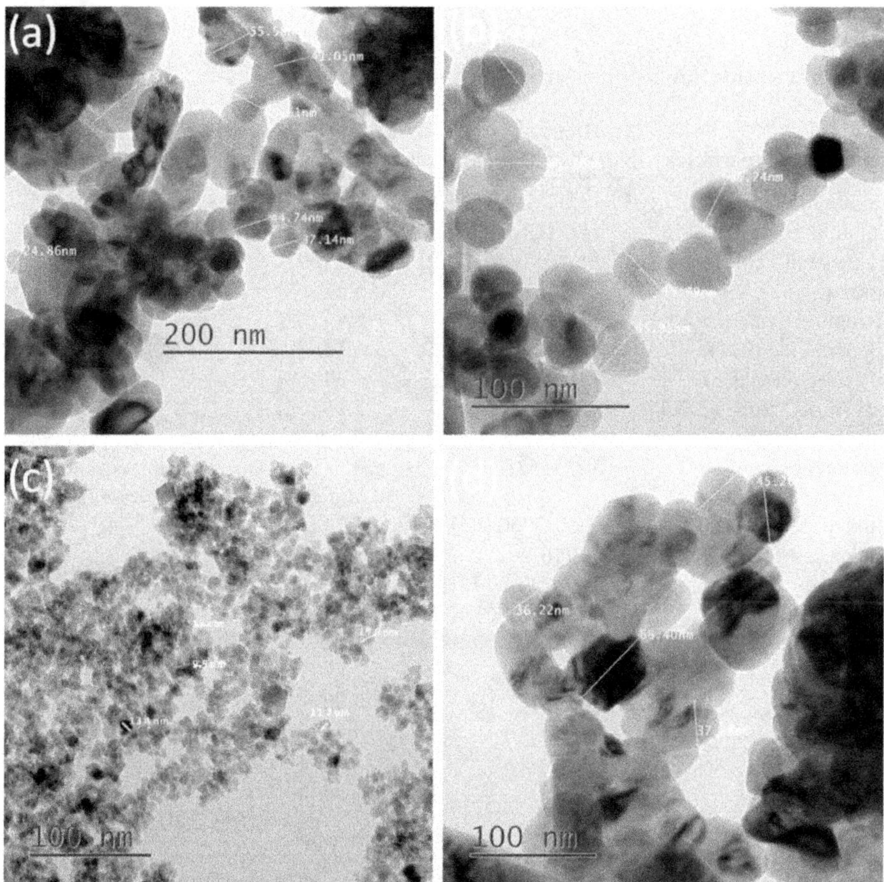

FIGURE 4.5 Nanoparticle characterization using TEM: (a) CuO, (b) ZnO, (c) Fe_2O_3 and (d) Al_2O_3. (From Ref. [63].)

nano-cutting fluids. These properties are sufficient to study the machinability behaviour of a material under nanofluid conditions.

The main aim of the addition of nanoparticles to a base coolant is to improve its thermal conductivity and thus improve heat transfer phenomena from the cutting zone (higher temperature) to open surroundings (low temperature). A list of the thermal conductivities of different nanoparticles is reported in Table 4.1. Sharmin et al. [58] observed that the thermal conductivity of nanofluids rose linearly with increasing volume concentrations of 0.2, 0.3, 0.4 and 0.5% with increase in nanofluid temperature from 60 to 90°C. Thermal conductivity was also observed to increase as the volume of CNTs increased. Andhare et al. [66] discovered a similar pattern. Das et al. [63] compared the thermal conductivity of CuO, ZnO, Fe_2O_3 and Al_2O_3 nanofluids with increasing temperature, as shown in Figure 4.6. The CuO based nanofluid has the greatest thermal conductivity in comparison to ZnO, Fe_2O_3 and Al_2O_3 nanofluids.

TABLE 4.1
Thermal conductivity and density of nanoparticles

Types of nanoparticles	Thermal conductivity (W/mK)	Density (g/cm^3)
Aluminium oxide (Al_2O_3)	40.0 [67, 68]	3.97 [68]
Titanium oxide (TiO_2)	11.7 [68]	4.23 [68]
Magnesium oxide (MgO)	54.9 [67] 48.4 [69]	2.9 [69]
Zinc oxide (ZnO)	29.0 [67]	5.61 [68]
Zirconium oxide (ZrO_2)	2 [68]	5.89 [68]
Copper oxide (CuO)	32.9 [67]	6.5 [68]
Iron (III) oxide (Fe_2O_3)	7 [68]	5.34 [68]
Iron (II, III) oxide (Fe_3O_4)	17.65 [68]	5.18 [68]
Silicon oxide (SiO_2)	7.6 [68]	2.4 [68]
Diamond	3000 [67]	3.5 [www.americanelements.com/diamond-nanoparticles-7782-40-3]
Multi-wall carbon nanotubes	2000-3000 [67]	1.74 [70]
Silicon carbide	490 [67]	3.21 [71]
Graphene	2000-5000 [72]	2-2.5 [73]
Graphite	2000-2200 [74]	2-2.5 [73]
Silver	429 [67]	7.9 [75]
Gold	315 [67]	19.3 [76]
Copper	398 [67]	8.96 [www.americanelements.com/copper-nanoparticles-7440-50-8]
Aluminium	247 [67]	2.7 [www.ssnano.com/inc/sdetail/aluminum_nanoparticles__nanopowder___al__99_9___100_130_nm_/12257]
Silicon	148 [67]	2.0 [https://nanocomposix.com/pages/silica-physical-properties#:~:text=The%20density%20of%20silica%20nanoparticles,atomic%20structure%20of%20bulk%20glass]

Kumar et al. [77] prepared water based Al_2O_3 and TiO_2 nanofluids and tested their thermal conductivity for three different weight % concentrations (0.005, 0.01 and 0.05). The thermal conductivity of both nanofluids increases with the weight % concentration of nanoparticles. Al_2O_3 nanofluid has better thermal conductivity than TiO2 nanofluid at each concentration because Al_2O_3 powder has higher thermal conductivity than TiO_2 powder. Padmini et al. [78] also estimated the thermal conductivity of vegetable oil based nanofluids. Various nanofluid samples were created by dispersing nanomolybdenum disulphide ($nMoS_2$) in coconut oil, sesame oil and canola oil with various nanoparticle volume concentrations (0.25, 0.5, 0.75 and 1%). The thermal conductivity of all prepared nano-cutting fluids was growing with leading concentrations. Coconut oil based MoS_2 nanofluid has the highest thermal conductivity in comparison to the others. Thakur et al. [79] measured the thermal conductivity of SiC nanofluids with different concentrations (0.5, 1 and 1.5%) at ambient

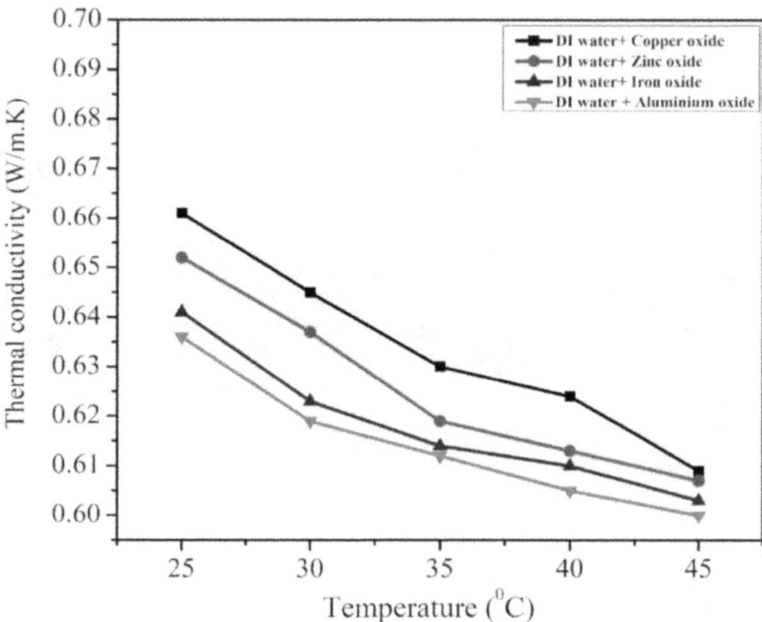

FIGURE 4.6 Comparison of the thermal conductivities of different nanofluids with increasing temperature. (From Ref. [63].)

temperature. The thermal conductivity of SiC nanofluids was raised as the particle content increased. The greatest coefficient of thermal conductivity was determined to be 0.7598 W/mK at 1.5 weight % concentration.

Further, viscosity is another index to show the behaviour of a nanofluid. Viscosity is an indicator of the shear strains imposed by one layer on the next layer to prevent flow. Das et al. [63] conducted viscosity tests of three distinct nanofluids (Al_2O_3, Fe_2O_3 and CuO). Al_2O_3 has the highest viscosity among the three, while CuO has the lowest. All nanofluids showed decreasing viscosity as temperature increased. Padmini et al. [78] executed kinematic viscosity tests of different vegetable oil based MoS_2 nanofluids. Coconut oil based nanofluids have the highest viscosity in comparison to sesame oil and canola oil based nanofluids. Also, for each nanofluid, the viscosity was improving with increasing particle concentrations, while it deteriorated with rising temperature. Kumar et al. [77] calculated the viscosity of water based Al_2O_3 and TiO_2 nanofluids at various concentrations. The viscosity of both nanofluids increased with particle concentration, although the improvement was minimal. When compared to the base fluid, the maximum increase in viscosity of Al_2O_3 nanofluid is 0.031% at the highest concentration (0.05% weight concentration), but the maximum improvement in viscosity of TiO_2 nanofluid is 0.029% at the highest concentration.

Wettability is another important characteristic of a nanofluid, and it refers to the ability of a liquid to spread over a surface and form a thin, continuous film. The addition of nanoparticles can alter the wettability of the fluid by changing the surface properties of the nanoparticles, such as their size, shape and surface charge [80]. When

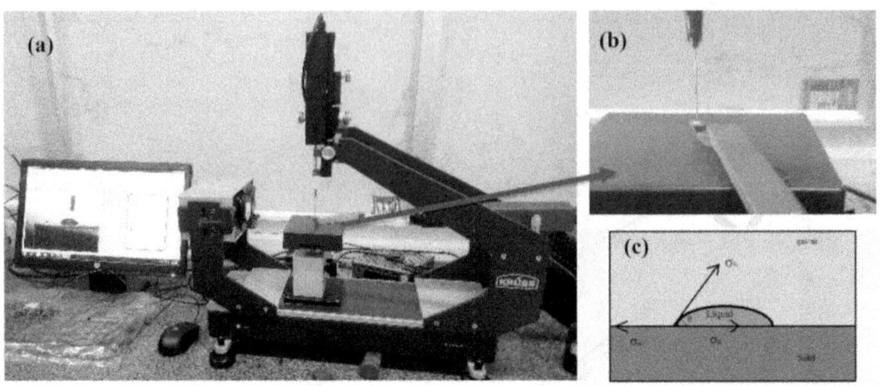

FIGURE 4.7 (a) The setup for measuring contact angles, (b) a close-up view of a dropper and a carbide tool and (c) a schematic of a liquid droplet on a solid surface. (From Ref. [81].)

nanoparticles are well dispersed in a fluid, they can enhance the wettability of the fluid by increasing its surface area and improving its wetting properties. This can lead to increased stability of the nanofluid by reducing the tendency of the nanoparticles to aggregate or settle out of the solution. On the other hand, if the nanoparticles are poorly dispersed in the fluid, they can decrease the wettability of the fluid, leading to instability and agglomeration of the nanoparticles [82]. The wettability of a nano-cutting fluid over the tool surface may improve heat extraction from the hot tool surface. The contact angle between the solid surface and the droplet may be used to assess the wettability properties of any cutting fluid. The typical arrangement for determining the macroscopic contact angle between a fluid droplet and the surface of a cemented carbide tool insert is shown in Figure 4.7 [81].

The inclusion of nanoparticles considerably affects the contact angle (wettability index parameter) of a nano-cutting fluid, as seen in Figure 4.8. From 0 to 1.5% nanoparticle concentrations, the contact angle first decreases for all nanofluid samples, and it subsequently rises for higher concentrations. Maximum wetting area per unit liquid volume was obtained by measuring the minimum contact angles for Al–GnP and alumina nanofluids to be 38.9° (at 1.0% volume concentration) and 41.9° (at 1.0% volume concentration), respectively. In contrast to the contact angles tested for each nanofluid, the base fluid was found to have a substantially larger contact angle of 54.9°. Therefore, compared to the base fluid, it has better heat extraction and lubricating capabilities [81]. Using the macroscopic contact angle approach, the wettability property of 1% of alumina dispersed nano-cutting fluid on a carbide tool-tip was measured by Khandekar et al. [83]. It was reported that the wetting area per unit volume rises for nanodroplets (10 microlitres) in comparison to dry and flood cooling. So, in comparison to the usual cutting fluid, it improves the lubricating and heat removal qualities.

The density of a nanofluid can vary depending on the type and concentration of nanoparticles and the base fluid used. When nanoparticles are added to a base fluid, the density of the resulting nanofluid is typically higher than that of the base fluid,

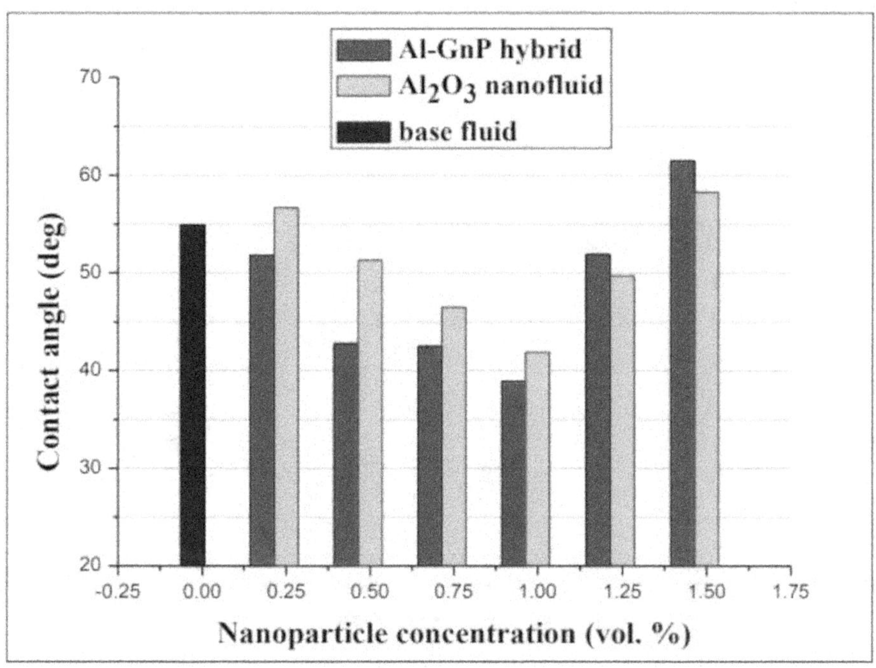

FIGURE 4.8 Effects of nanoparticle concentration on contact angle. (From Ref. [81].)

due to the additional mass of the nanoparticles. The density of a nanofluid can be calculated using the following equation [77]

$$\rho_{nanofluid} = \phi.\rho_{nanoparticles} + (1-\phi)\rho_{basefluid} \quad 4.1$$

where "$\rho_{nanofluid}$" is the density of the nanofluid, "$\rho_{nanoparticles}$" is the density of the nanoparticles, "$\rho_{basefluid}$" is the density of the base fluid an "ϕ" denotes the concentration (by volume) of the nanofluid to be synthesized.

4.2.3 Stability

Nanofluids eventually lose their heat transfer capability when nanoparticles settle over time. Thus, it is critical to address nanofluid stability since it may alter the thermophysical qualities necessary for application [62]. The stability of nanofluids may be divided into three categories: dispersion, kinetic and chemical stability. Dispersion stability is concerned with avoiding nanoparticle aggregation in a fluid. Kinetic stability is achieved by impeding gravitational sedimentation via Brownian diffusion of nanoparticles. Chemical stability is determined by the reactivity of nanoparticles with the base fluid, which is unsuitable for cooling purposes. Thus, the two most important features of nanofluid stability are avoiding nanoparticle aggregation and sedimentation. The DLVO (Derjaguin, Landau, Verwey, Overbeek) hypothesis is the best method for estimating the dispersion stability of base fluids [62].

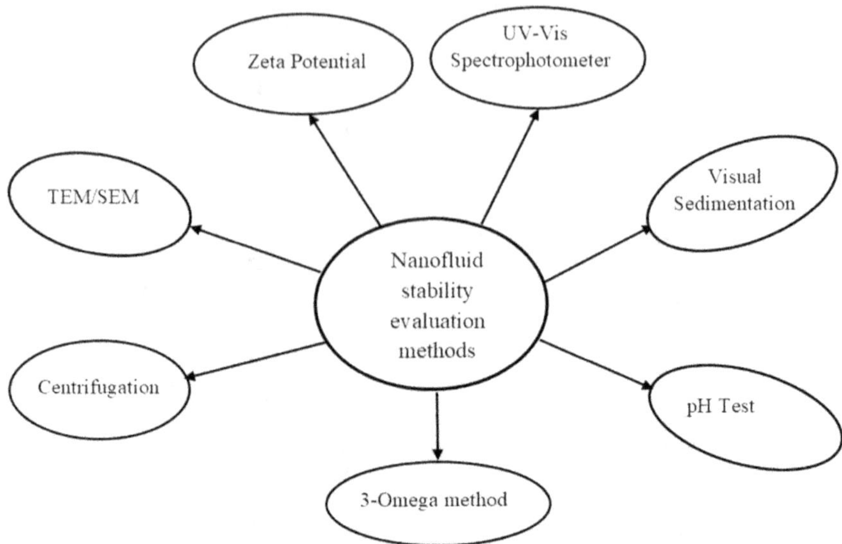

FIGURE 4.9 Common methods for assessing the stability of nanofluids.

A recent modification in the DLVO theory provided a clear description of the constancy of nanofluids. This theory explained the concept of the van der Waals force of attraction and electrostatic repulsion [84]. DLVO theory is concerned with the van der Waals force of attraction and electrostatic repulsion caused by the development of electrical double layers. Another essential characteristic for describing the stability of nanofluids is the formation of electrical double layers. Van der Waals attraction causes nanoparticles in base fluids to aggregate, but the ion diffusion process prevents this from happening. Exploring techniques and applying nanofluids in the actual world need the proper evolution of nanofluid stability. Some popular methods used to assess the stability of nanofluids are shown in Figure 4.9.

The zeta potential reflects the potential difference between the fixed layer of a base fluid affixed to the particles and the dispersion medium. It produces an electrostatic field that in turn generates an electric surface charge that affects the ions in the liquid's main body. The electrostatic field produced by the thermal motion of the ion serves as a screen for the electric surface charge. This diffuse layer of screening generates an electric charge that is equal in magnitude and opposite in polarity to the net surface charge. As a result, the entire setup is electrically neutral; this is conceptually shown in Figure 4.10 [85]. Above 60 mV, nanofluids have outstanding stability, above 30 mV, they are physically stable, below 20 mV, they have low stability and, below 5 mV, they are highly unstable and ready to clump together.

Zeta potential analysis was carried out by several researchers to assess the stability of various nanofluids. Arifuddin et al. [86] carried out the zeta potential test of hybrid (alumina and titanium oxide) nano-cutting fluid. The nano-cutting fluid of 0.00001% volume concentration has a 37.6 mV zeta potential. The hybrid Al_2O_3–TiO_2 nanofluid's zeta potential value shows that the concentration of the nanofluid has

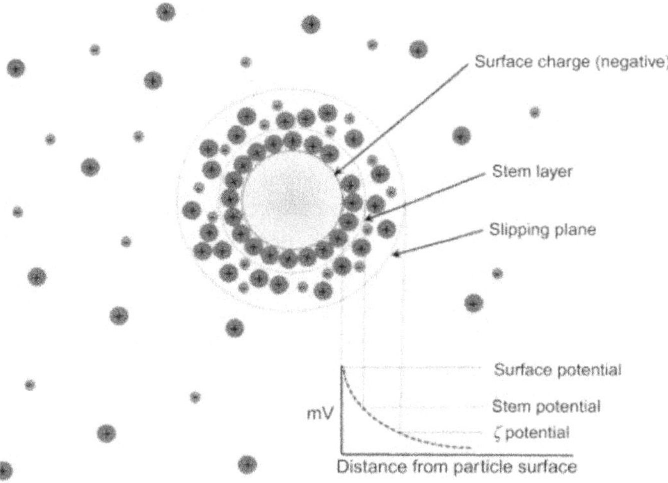

FIGURE 4.10 Concept of zeta potential. (From Ref. [85].)

a very high correlation with its zeta potential value. The nanofluid is incredibly stable, with a zeta value of 64.2 mV and a volume concentration of 0.001%. Alumina and titanium oxide based hybrid nanofluids with higher concentrations are therefore more appropriate for long-term usage than those with lower concentrations. According to Gaurav et al. [87], the zeta potential decreases with increasing nanoparticle concentration, especially at 0.9% concentration. As the zeta potential approaches the threshold value of 15 mV, the nanofluid becomes unstable. The jojoba nanofluid's results for the zeta potential were in the same ballpark as those seen for other vegetable nanofluids. For instance, rapeseed oil with MoS_2 dispersed nanofluid displayed zeta potentials of 28.5 and 33.5 mV for nanoparticle concentrations of 0.5% and 1.0%, respectively [88].

UV–Vis spectroscopy is a commonly used analytical technique for characterizing nanoparticles in solution [77]. The technique relies on measuring the absorbance of light by the sample at different wavelengths, which is related to the concentration and size of the particles in the solution. UV–Vis spectroscopy can be used to determine the stability of nanoparticles by monitoring the changes in their absorbance spectra over time. If the absorbance spectrum of the nanoparticles remains constant over time, it suggests that the nanoparticles are stable. However, if there are changes in the absorbance spectrum, such as a shift in the peak position or a decrease in the intensity of the extinction bands, it suggests that the nanoparticles are undergoing aggregation or precipitation, which indicates instability. The stability of a nanofluid can be effectively estimated by the UV–Vis spectroscopy method when nanoparticles have extinction bands between 190–1100 nm wavelengths [62].

Arifuddin et al. [86] also conducted a UV–Vis spectrophotometer test to check the stability of a hybrid (Al_2O_3 and TiO_2) nanofluid for cutting fluid application. With sonication times of 0, 30 and 60 min, the hybrid nano-cutting fluid (0.001% volume concentration) had a better absorption ratio than the fluid with a 120 min sonication time. After 90 min of sonication, the hybrid nano-cutting fluids showed the greatest absorption ratio. The most stable hybrid nano-cutting fluid was created after 90 min of sonication. Kumar et al. [77] conducted a UV–Vis spectrophotometer test for two different nanofluids (Al_2O_3 and TiO_2 based) at different nanofluid concentrations (0.005%, 0.01% and 0.05% by weight). The absorbance of the nanofluids ranged between 200 and 900 nm. The absorbance of Al_2O_3 nanofluid is greater than that of TiO_2 nanofluid, indicating that more agglomerated nanoparticles are present in Al_2O_3 nanofluid than in TiO_2 nanofluid. The absorbance of suspended nanoparticles rises as the weight % concentration of the nanofluids rises. So, when weight concentration increases, nanoparticle stability decreases. Additionally, it can be inferred from this research that the colloidal stability of the nanofluids is maintained at each weight % concentration of both nanofluids.

Visual sedimentation is one of the easiest and most common techniques to estimate the stability behaviour of a nanofluid. Higher mass concentrations inside the nanofluid enhance the density and viscosity at room temperature and, as a consequence, aggregation is avoided at comparatively higher concentrations, thus nanoparticle suspension stability improves [87]. Arifuddin et al. [86] conducted sedimentation tests for four different concentrations of hybrid (Al_2O_3 and TiO_2) nanofluid (1%, 2%, 3% and 4% by volume) for a period of 30 days. No obvious nanoparticle sedimentation in the nanofluid for any concentration was noticed due to the fact that nanoparticle aggregation had not yet occurred. Therefore, this confirmed that the nanofluids are stable immediately after preparation.

One of the most straightforward and popular ways to determine a nanofluid's stability is to look at its pH value [89]. A nanofluid's pH level can offer crucial information on its thermal conductivity and stability [90]. The pH value of nanofluids exposes their homogeneity, with the nanofluid with the highest pH value being regarded as the most homogeneous [91]. When a nanofluid's pH value is lower than the isoelectric point, it is said to be stable [63]. Al_2O_3 and ZnO nanoparticles have isoelectric points of 9.2 and 9.8, respectively [89]. While manufactured Al_2O_3 and ZnO nanofluids were calculated to have pH values of 7.7 and 7.5, respectively, LRT 30 oil was determined to have a pH value of 7.5. Both nanofluids are stable, though, because the resulting pH value for each is considerably below the corresponding isoelectric point. It may be argued that the ZnO nanofluid was comparatively more stable than the Al_2O_3 nanofluid because the pH value of ZnO was relatively distant from the isoelectric point. Yildirim et al. [90] evaluated the pH and viscosity of different mono and hybrid nanofluids, and the results are graphically shown in Figure 4.11. The base fluid had the lowest viscosity (40.3 mPa.s) and pH value (5.61). The pH and viscosity of both mono and hybrid nanofluids were raised by the addition of nanoparticles to the base fluid. MoS_2 nanofluid had a viscosity value of 48.3 mPa.s and a pH value of 7.9, whereas Al_2O_3:MoS_2 (1:1) nanofluid had the greatest value. In other words, as compared to the base liquid, the pH value of the Al_2O_3:MoS_2 (1:1) hybrid nanofluid had been raised by 40.82%, and the viscosity value was enhanced by 19.85%. Al_2O_3–MoS_2 type hybrid nanofluids were found to have greater viscosity and pH values than other hybrid nanofluids among the manufactured nanofluids.

The 3-omega technique measures nanofluid stability by measuring thermal conductivity fluctuations caused by nanoparticles settling [62, 92]. Centrifugation is

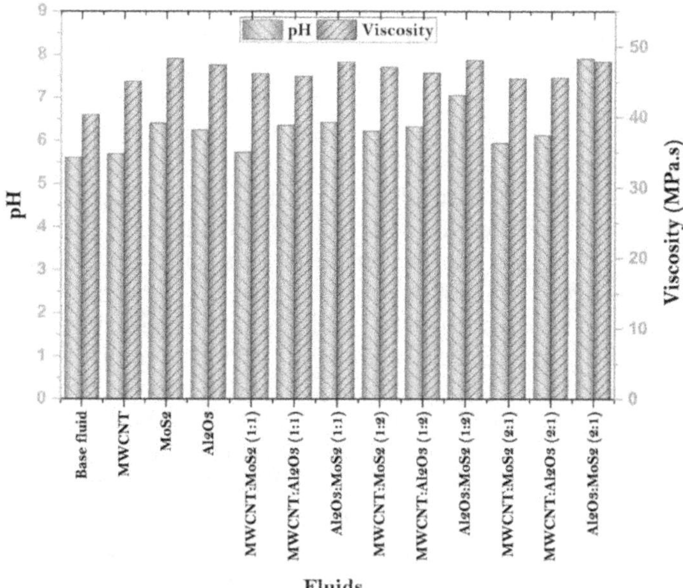

FIGURE 4.11 pH and viscosity results of different mono and hybrid nanofluids. (From Ref. [90].)

commonly used to assess the stability of nanofluids. This approach is employed for assessing the stability of Ag–ethanol nanofluids made using microwave synthesis with polyvinyl alcohol as a stabilizer. The nanofluids were stable for ten hours at 3000 rpm centrifugation [62]. Using this approach, Ditsch et al. [93] investigated the clustering and stability of polymer coated magnetic nanoparticles.

Furthermore, TEM is used to investigate the dispersion of hybrid nanoparticles in liquid form, therefore verifying their stability [86]. Particle clustering may be seen by looking at high-resolution optical micrographs of dried nanofluid samples acquired through TEM or SEM. It is possible to identify and analyze the microstructural changes brought on by aggregation or clustering, as well as the overlap of nanoparticles in the distribution. While clustered or grouped nanoparticles in the specimen sample suggest poorly distributed unstable nanofluids, homogeneous distribution of particles implies well-dispersed stable nanofluids. Using TEM, Khaleduzzama et al. [94] depicted the microstructure and particle dispersion of alumina–water nanofluids. Within the range of 100 nm, they saw a respectable dispersal of nanoparticles.

4.3 MACHINING PERFORMANCE USING NANO-CUTTING FLUIDS

Nano-cutting fluids have recently grabbed the attention of researchers and engineers due to their efficient cooling capabilities in machining processes. A large amount of research has been conducted on increasing the thermal characteristics of various nano-cutting fluids, and it is obvious from the literature that the inclusion of nanoparticles in any base fluid enhances thermal conductivity, heat carrying capacity and viscosity. Improving these features of the base cutting fluid considerably enhanced tool life, surface quality and surface integrity, and encourages sustainable machining [95–97]. The key research findings on machining performance using nano-cutting fluid are discussed briefly as follows:

Das et al. [98] investigated the application of emulsion water based oil with/without aluminium oxide nanoparticles via MQL in the hard turning of AISI 4340 alloy steel. The research also looked at the machining process under air-cooling conditions. The obtained findings demonstrated that the cutting forces under MQL conditions employing nanofluid were the lowest, followed by air cooling and without nanofluid-added emulsion. Furthermore, the stability of cutting forces under Al_2O_3 nano-cutting fluid can be clearly detected, demonstrating the Al_2O_3 nanofluid's improved lubricating and cooling capabilities. Sharma et al. [99] investigated the effect of CNT nanoparticles on the hard turning of AISI D2 steel under MQL conditions, and they discovered that the thermal conductivity of the CNT nanofluid was improved when compared to the base fluid without nanoparticles. As a result, the surface quality increased and tool wear was decreased. Gupta et al. [100] examined the performance of pure vegetable oil and Al_2O_3 vegetable oil based nanofluid in AISI 4130 hard turning for sustainable manufacturing. When compared to pure vegetable oil, the authors discovered that turning performance improved with Al_2O_3 nanofluid and surface quality is improved by roughly 27.3%. Yıldırım [60] evaluated the results of hard turning AISI 420 steels with CVD coated carbide inserts under nanofluid MQL (NFMQL) and cryogenic cooling conditions. In this investigation, 0.5% volume concentration of GnP was mixed with Fuchs Plantocut 10 SR oil to synthesize a nanofluid, and this was used as a coolant via MQL in the hard turning of AISI 420 steel. The performance of the nanofluid was compared with that of liquid N_2 (LN_2). NFMQL outperformed cryogenic chilling conditions in terms of surface roughness and surface topology. At a constant feed rate of 0.1 mm/rev with cutting speeds of 75 m/min, 100 m/min and 125 m/min, the nanofluid improves surface roughness by 20.78%, 47.38% and 14.66%, respectively, as compared to cryogenic cooling. Figure 4.12 indicates the difference in surface roughness obtained under cryogenic and NFMQL conditions with different cutting parameters. The least surface roughness in nanofluid was obtained at the lowest feed (0.1 mm/rev) with highest speed (125 m/min) turning conditions, while the least surface roughness under cryo-turning was obtained at the lowest feed (0.1 mm/rev) with moderate speed (100 m/min) turning conditions. Furthermore, in both nanofluid and cryogenic cooling scenarios, SEM pictures of the tool-tip (Figure 4.13 (a) and Figure 4.13 (b)) were taken at a speed of 75 m/min. Cryogenic cooling caused less wear than nanofluid cooling, however, edge fracture occurred in both cooling settings. BUE was seen under both scenarios. Coating peeling was seen under LN_2 conditions. Adhesion was also seen on the tool's surface under both cooling conditions. Chip adhesion was clearly seen in the nanofluid condition.

Under NFMQL conditions, Duc et al. [101] applied hard turning of heat-treated 90CrSi steel using tungsten carbide inserts. Nanofluids were synthesized by mixing Al_2O_3 and MoS_2 nanosized powder in soybean oil and water based emulsions,

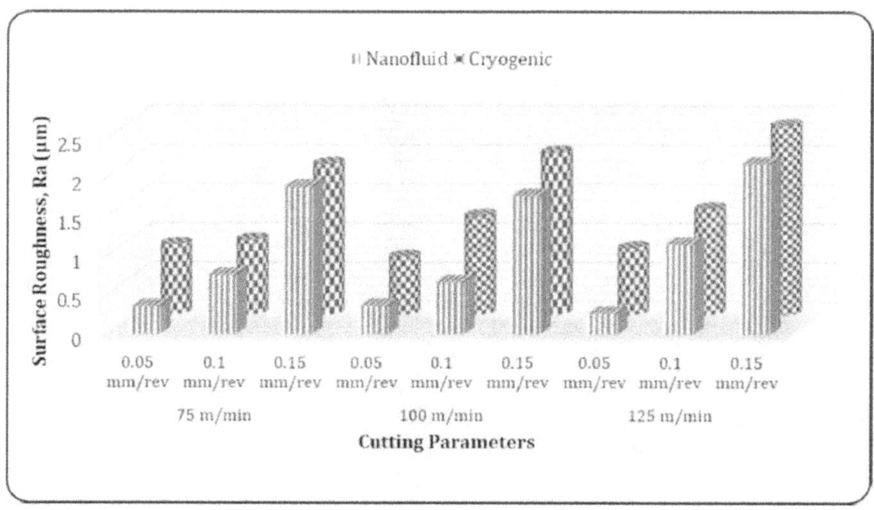

FIGURE 4.12 Comparison of effects of cutting parameters on surface roughness under NFMQL and LN_2 cryo-cooling. (From Ref. [60].)

respectively. MoS_2 nanofluid has a greater impact on lowering machining forces and feed force than Al_2O_3 nanofluid, but because thrust force has the largest influence on dimensional precision, Al_2O_3 nanofluid can be employed in this situation. Liew et al. [102] demonstrated a hard turning process on heat-treated AISI D2 steel with PVD coated carbide inserts in a nanofluid and pure deionized water environment at a constant flow rate of 200 ml/min. Carbon nanofibre (CNF) nanofluid is created by mixing carbon and deionized water with the surfactant gum arabic. It was discovered that as cutting speed increases with a greater feed rate, tool wear increases but surface smoothness improves. This is due to CNF having stronger thermal conductivity and better tribological characteristics than pure deionized water. Patole and Kulkarni [103] investigated the hard turning of AISI 4340 steel under MQL coupled MWCNT nanofluid. According to the findings, the feed rate was the most significant variable in producing lowered surface roughness under MQL utilizing nanofluid, followed by the depth of cut, while cutting speed was the factor with the least significance. Sharma et al. [104] discovered that, compared to standard cutting fluid and dry circumstances, employing larger concentrations of nanoparticles in nanofluid increased surface quality and lubricating action in machining. During machining with the nanofluid surroundings, minimizing trends of frictions, forces, power usage, flank wear and temperature were also noticed. Mode of lubrication, nano-element size, nano-element concentration, nozzle orientation, air pressure and spraying distance were all significant criteria that influenced total machining performance. Khalil et al. [105] evaluated the impact of employing Al_2O_3 nano-lubricant while turning AISI 1050 grade steel with a coated carbide cutting tool and discovered decreased tool wear and extended tool life. Khajehzadeh et al. [106] conducted an experimental study on water based TiO_2 nanofluid utilizing hardened AISI 4140 steel and discovered that increasing

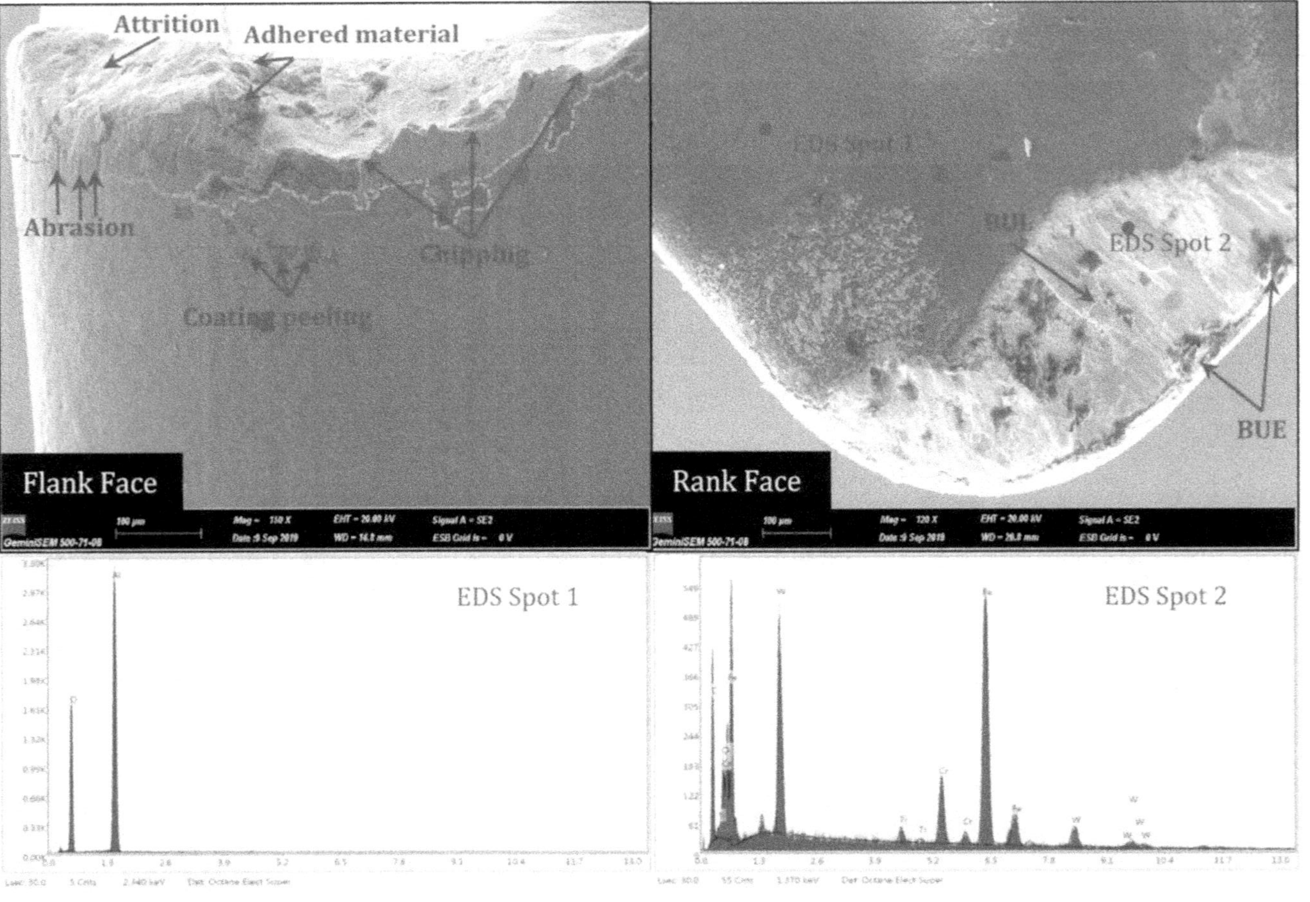

FIGURE 4.13 SEM and EDS study of a worn tool acquired at 75 m/min cutting speed under (a) nanofluid cutting environment and (b) LN_2 cutting environment. (From Ref. [60].)

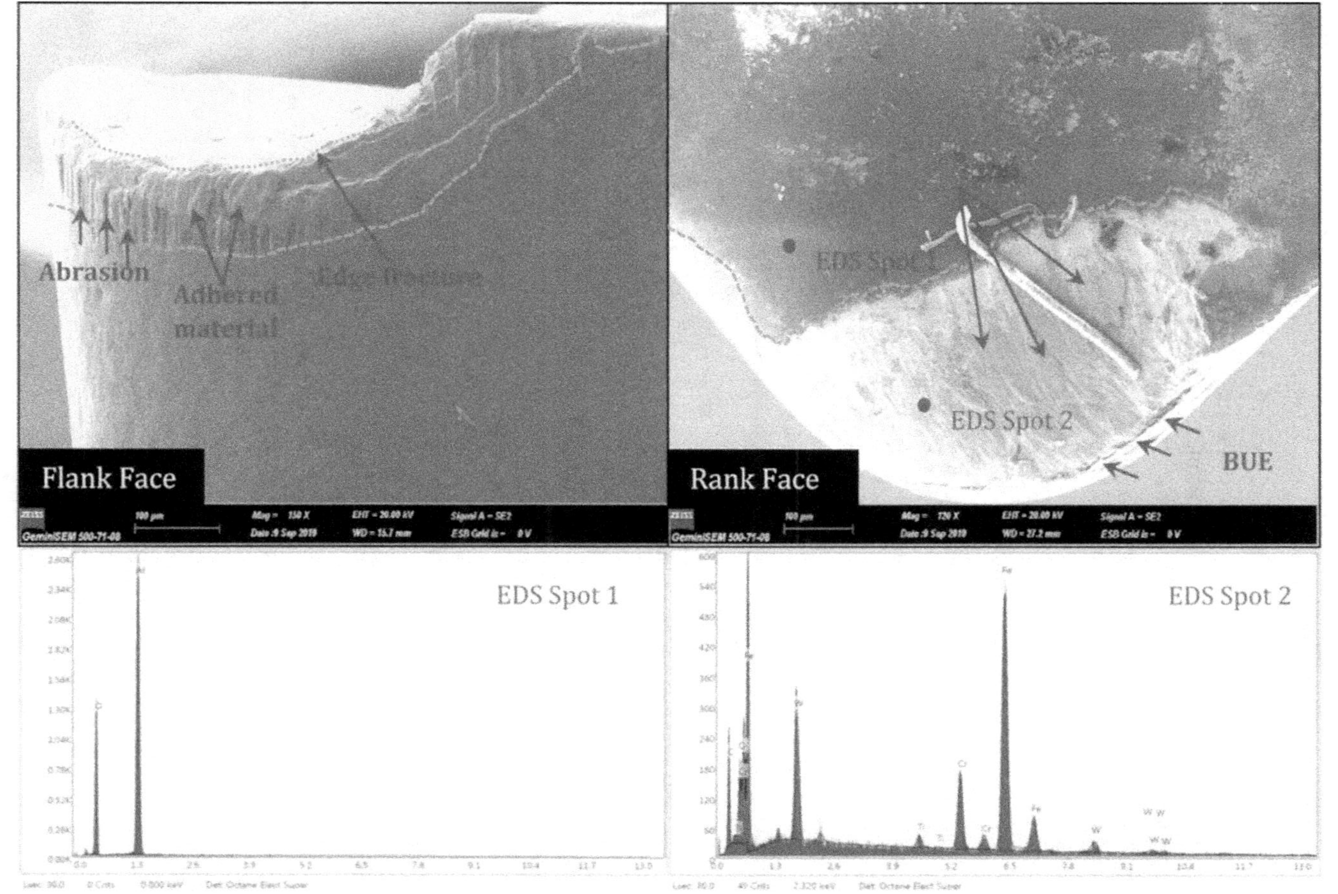

FIGURE 4.13 (Continued)

nanoparticle concentration from 3% to 5% or decreasing nanoparticle size from 50 to 10 nm resulted in decreased rates of cutting tool flank wear. It can also increase a component's machining performance and total production cost. Das et al. [107] implemented Al_2O_3 based nanofluid in machining hardened AISI 4340 steel and discovered a significant improvement in tool wear when compared to water-soluble and compressed air coolants. The relative changes in forces with flank wear have been examined for all three cutting coolant settings (compressed air, water-soluble coolant, and nanofluid). Figure 4.14 depicts the effect of flank wear on machining forces. The intensity of forces is substantially influenced by flank wear. In comparison to other environments, the nanofluid machining environment has the narrowest range of cutting forces.

In a further study, Das et al. [59] utilized three different nanofluids (CuO, Al_2O_3 and Fe_2O_3) via MQL in hard turning. CuO nanofluid exhibited superior results among

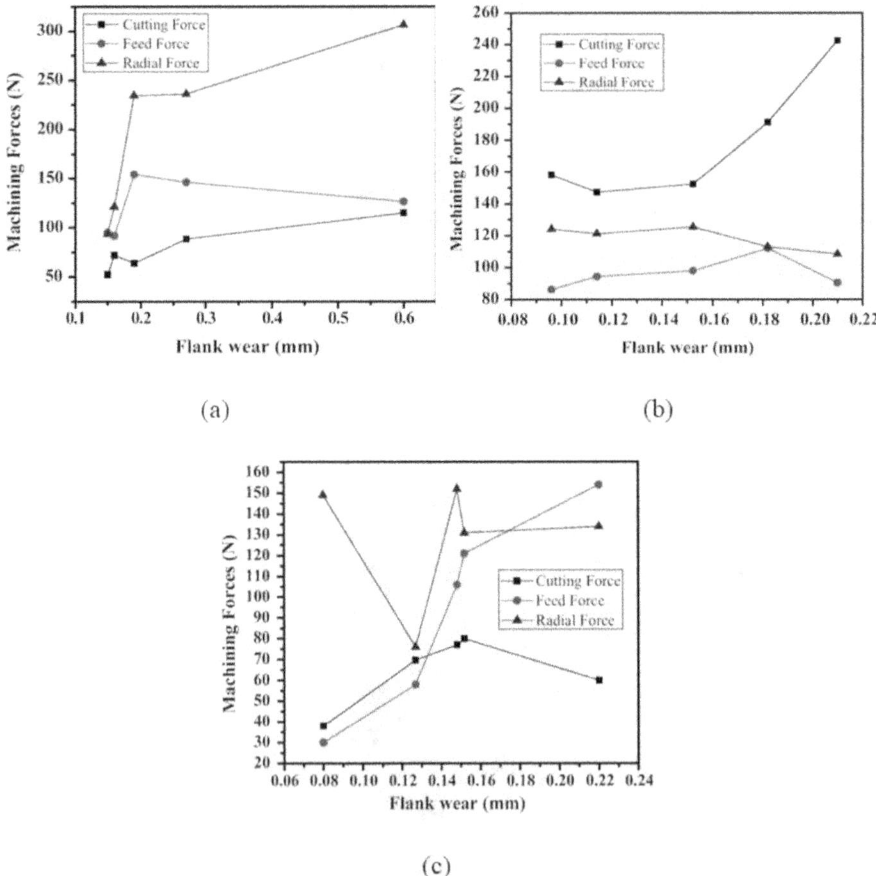

FIGURE 4.14 Effects of flank wear progression on turning forces under various cutting environments: (a) compressed air, (b) water-soluble coolant and (c) nanofluid. (From Ref. [98].)

these nanofluids as the least flank wear and lower surface roughness were noticed. Ibrahim et al. [108] implemented MQL and ZnO NFMQL coolants to perform hard turning tests on hardened AISI D3 steel. They evaluated the performance of a 0.1% weight concentration of rice bran oil based ZnO nanofluid as compared to that of standalone MQL, dry and flood cooling. The NFMQL produced the lowest cutting force. In comparison to dry and flooded machining, NFMQL machining outperformed the optimum lubrication-cooling approach due to lower cutting force, improved surface morphology and less tool wear.

Tun et al. [109] performed turning experiments on 90CrSi low alloy steel using MoS_2 NFMQL and observed that in terms of surface topography and surface roughness, MQL with MoS_2 nano-additions outperformed pure MQL. White layer, burn markings and surface deformation were also decreased as a result of the increased lubricating and cooling performance of the nanofluid. Prasad and Srikant [110] observed that increasing the nano-graphite concentration of the cutting fluid enhanced both thermal and machining performance. When hardened AISI 1040 steel was machined with uncoated cemented carbide and HSS cutting inserts, higher MQL flow rates resulted in increased machining performance. Liew et al. [111] investigated the machining performance of AISI D2 steel under CNF nanofluid and deionized water. The results showed that employing CNF nanofluids instead of deionized water improved surface properties and tool life due to the CNF's higher tribological and thermal conductivity. Sayuti et al. [112] conducted a turning operation on hard-to-cut AISI 4140 steel with coated carbide tools in order to investigate surface roughness, oil depletion and tool wear. Using the MQL technique, nanofluid was generated in this experiment by adding SiO_2 to the mineral oil used as a coolant. According to the findings, the lowest tool wear was obtained at 2 bar air stream pressure, 60° nozzle angle and a weight concentration of 0.5% nanoparticles in the base fluid. Duc et al. [113] studied the tool wear, turning force and surface roughness in milling 60Si2Mn hard-to-cut steels using a coated carbide tool under three different concentrations (0.5%, 1.0% and 1.5 % by weight) of soybean oil based alumina nanofluids. When the concentration of the nanofluid is increased from 0.5% to 1.55%, the tool's life increases from 80 to 115 min, i.e. 43.8% increment. According to the findings, a soybean based nanoparticle dispersion of 1.0–1.5% may be used for hardened part milling to achieve an excellent surface finish while increasing tool life and reducing waste caused by nanoparticle precipitation. Gajrani et al. [114] employed a carbide insert in a range of cutting conditions, including using mineral oil, green cutting fluid (GCF) and a hybrid nano green cutting fluid (HNGCF), during a hardened part turning technique of AISI H13 steel. CaF_2 and MoS_2 were dispersed in GCF based on vegetable oil to make HNGCF. The HNGCF0.3M (0.3% concentration of MoS_2 based HNGCF) environment was found to have reduced cutting tool wear and workpiece material adherence compared to other cutting settings. HNGCF0.3M generates the best surface quality in all cutting settings as compared to mineral oil, GCF and HNGCF0.3C (HNGCF based on a 0.3% concentration of CaF_2), enhancing surface finish by 37.21 %, 12.32 % and 13.23 %, respectively.

Chavan and Sargade [115] investigated turning performance of AISI 52100 steel under various cooling cutting conditions, including dry, minimum quantity cooling and lubrication (MQCL), compressed cold air, and hybrid nanofluid MQCL. Al_2O_3 and

MWCNT are dispersed in soyabean oil for hybrid nanofluid formation, and soyabean oil is employed under MQCL conditions. Under the hybrid nano-cutting fluid condition, 0.34 μm surface finish, 625 $Hv_{0.1}$ micro-hardness, 168 MPa residual stress and 0.9 μm white layer thickness were detected after the experiment. Minh et al. [116] executed milling operations on 60Si2Mn steel in various MQL and NFMQL environments. In this experiment, four types of cutting fluids are used: [soybean oil, soybean oil + nano Al_2O_3 (0.5% weight concentration), emulsion 5% coolant, emulsion 5% coolant + nano Al_2O_3 (0.5% weight concentration)]. The use of soybean oil based nanofluids in MQL hard milling was less effective than emulsion. As a consequence, emulsion based nanofluid outperformed soybean based nanofluid in terms of tool life. As a result, the utilization of soybean oil based nanofluids in MQL hard milling was less successful than emulsion. Duc et al. [117] carried out hardened part milling on Hardox 500 steel under MQCL conditions. Al_2O_3 and MoS_2 are distributed in rice bran oil to create a hybrid nanofluid. By raising the cutting speed under Al_2O_3/MoS_2 hybrid nanofluid MQCL conditions, the cutting speed may be increased from 80 to 140 m/min to get a high-quality surface finish. However, machining at a speed of 140 m/min is found to be the most economically and technologically efficient condition for hard milling. Uysal et al. [118] evaluated the influence of cutting parameters on initial tool wear and surface roughness in milling of AISI 420 martensitic stainless steel. The trials were carried out in three different conditions: dry, MQL with vegetable cutting fluid, and MQL with nanofluid. The MQL with nanofluid strategy exhibited improved performance relative to dry and MQL. The use of cutting fluid containing dispersed MoS_2 nanoparticles in MQL milling resulted in minimal tool wear and surface roughness, owing to the lubricating effect of MoS_2 nanoparticles.

4.4 MACHINING PERFORMANCE USING HYBRID NANO-CUTTING FLUIDS

Hybrid nanofluid is a mixture of two or more types of nanoparticles suspended in a base fluid. The nanoparticles can be of different materials, sizes and shapes. The properties of a hybrid nanofluid can be tuned by varying the composition and concentration of the nanoparticles. Hybrid nanofluids have shown improved thermal and electrical conductivity and enhanced lubrication properties compared to mono nanofluids. Hybrid nanofluids have been found to offer several benefits over conventional nanofluids [86]. Some of these benefits include:

- Enhanced thermal conductivity: Hybrid nanofluids have been shown to have higher thermal conductivity than conventional nanofluids, which can lead to better heat transfer in various applications.
- Improved lubrication properties: Hybrid nanofluids can improve the lubrication properties of base fluids, which can reduce friction and wear in moving mechanical components.
- Customizable properties: By varying the composition and concentration of the nanoparticles, the properties of hybrid nanofluids can be tailored to meet specific requirements for various applications.

- Increased stability: Hybrid nanofluids have been found to be more stable than conventional nanofluids, which can reduce the risk of nanoparticle agglomeration and sedimentation.
- Reduced cost: By using multiple types of nanoparticles, it is possible to achieve the desired properties of a hybrid nanofluid with a lower overall concentration of nanoparticles, reducing the cost of the nanofluid.

Overall, the use of hybrid nanofluids shows great promise for a wide range of applications, including in heat transfer, lubrication and electronics cooling, among others. Hybrid nanofluids have been proposed as a potential coolant in hard machining applications due to their enhanced thermal conductivity, which can improve the efficiency and quality of the machining process. The addition of nanoparticles to a base fluid can significantly increase its thermal conductivity and convective heat transfer coefficient.

Many studies have shown that hybrid nanofluids can improve the thermal performance of the cutting process by reducing the cutting temperature and improving heat transfer between the tool and workpiece. This can lead to improved surface quality, reduced tool wear and longer tool life. Sharma et al. [119] have developed a new type of cutting fluid by combining alumina based cutting fluid and MWCNT nanoparticles in volume concentrations of 0.25%, 0.75% and 1.25%. The researchers evaluated the thermophysical properties of the hybrid nanofluid and compared it to the base nanofluid, which was alumina-based. The addition of 1.25% volume concentration of Al-MWCNT and 1.25% volume concentration of alumina-based nanofluids resulted in enhancements of thermal conductivity by 11.13% and 9.85%, respectively. The utilization of a hybrid nanofluid consisting of Al-MWCNT at a volume concentration of 1.25% effectively reduced wear. The results also indicated that the use of the Al-MWCNT hybrid nanofluid led to a reduction in cutting force, thrust force, feed force, and surface roughness, with the lowest values recorded at 110.20 N, 85.48 N, 32.86 N and 0.953 mm, respectively. Overall, the findings suggest that the Al-MWCNT hybrid nanofluid is an effective cutting fluid that reduces wear and improves machining performance.

In another study, Sharma et al. [120] examined the impact of the amalgamation of two distinct nanofluids, namely alumina and molybdenum disulphide, in turning AISI 304 stainless steel. The present study involved the synthesis of a hybrid nanofluid through the amalgamation of alumina based nanofluid and molybdenum disulphide (MoS_2) nanoparticles, in a constant volumetric ratio of 90:10. The thermophysical properties of a prepared base fluid consisting of alumina nanofluid and a hybrid nanofluid with varying nanoparticle volume concentrations of 0.25%, 0.75% and 1.25% were evaluated at different temperatures. The utilisation of Al–MoS_2 hybrid nanofluid as a cutting fluid has resulted in a noteworthy decrease of 7.35%, 18.08%, 5.73% and 2.38% in cutting force (Fz), feed force (Fx), thrust force (Fy) and surface roughness (Ra), respectively, in comparison to Al_2O_3 mixed nanofluid. The application of Al–MoS_2 hybrid nanofluid as a cutting fluid has led to the attainment of the minimum values of 127.94 N, 88.92 N, 35.84 N and 1.396 μm, respectively, for cutting force, thrust force, feed force and surface roughness. A decrease of 2.4% in surface roughness was observed in comparison to cutting fluid enriched with alumina nanoparticles. The observed phenomenon can be attributed to the decrease in the coefficient of friction resulting from the presence of a Mo–S–P layer between the surfaces undergoing sliding. Figure 4.15 demonstrates a significant disparity in surface quality, indicating that the utilization of Al–MoS_2 hybrid nanofluid can result in a substantially enhanced surface. A high amount of sliding marks, pits, wear debris, worn out flakes and galling of surface are noticed during machining under dry, base

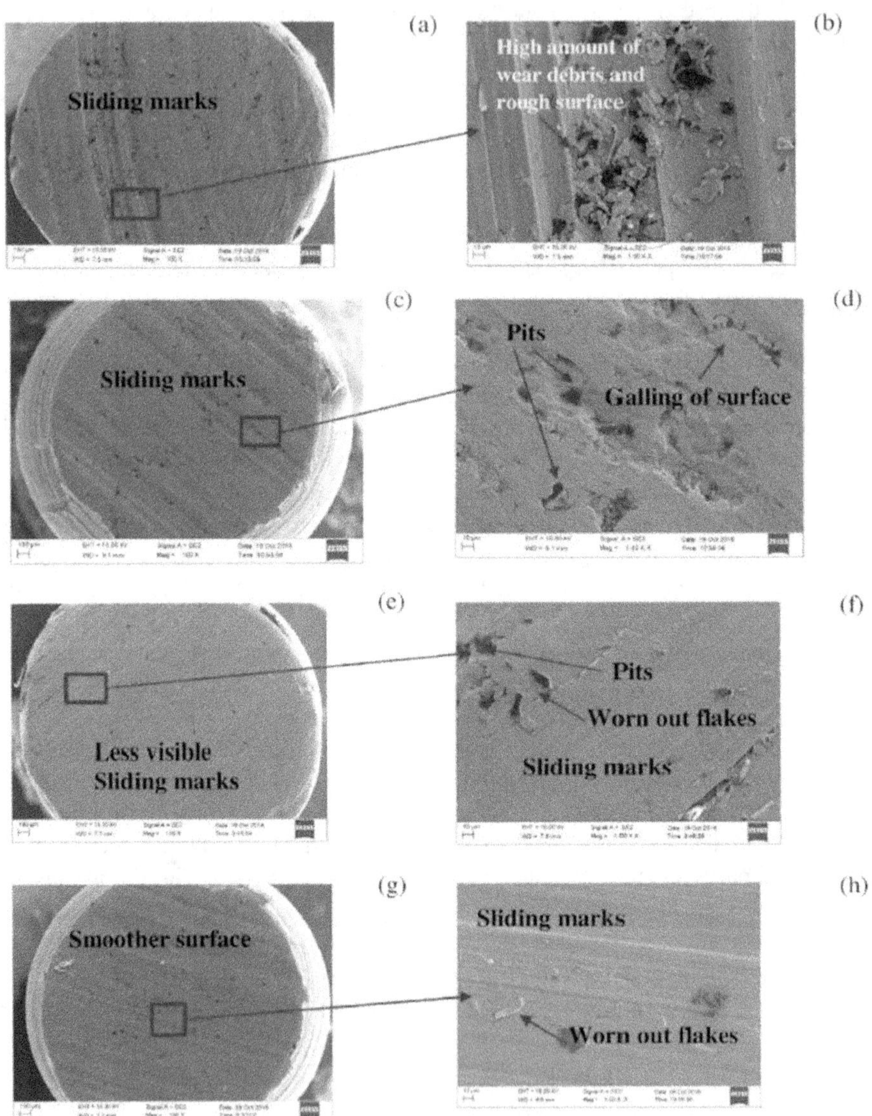

FIGURE 4.15 FESEM micrograph of pin wear under (a–b) dry, (c–d) base fluid, (e–f) Al_2O_3 and (g–h) Al–MoS_2 nanofluids in a pin-on-disc test. (From Ref. [120].)

fluid and Al_2O_3 nanofluid environments compared to the Al–MoS_2 hybrid nanofluid environment. This provides evidence of its superiority as a lubricant compared to both the base fluid and the alumina based nanofluid.

Gugulothu et al. [121] investigated the effect of hybrid nanofluid in the hard part turning process of AISI 1040 steel. Six hybrid nano-cutting fluids with different thermal conductivity, specific heat, viscosity and sedimentation concentrations were tested and the machining performance under them has been tested. It was observed that CNT/MoS_2 nanoparticles in sesame oil enhanced thermal conductivity, specific heat and viscosity significantly. Nanoparticle concentration in the base fluid of up to 2% by weight reduced the coefficient of friction prominently, and then it ascended slightly. MQL machining trials used different hybrid nano-cutting fluid concentrations, however, the 2% weight concentration CNT/MoS_2 hybrid nano-cutting fluid decreased cutting forces, temperature, surface roughness and tool flank wear the most. Minimal sedimentation was maintained in this concentration. Compared to dry machining and conventional cutting fluid, respectively, thrust force decreased by 22% and 11.2%, feed force by 28.3% and 13.8%, and main cutting force by 32% and 27.3% in CNT/MoS_2 hybrid nano-cutting fluid at 2% weight concentration. Compared to dry machining and conventional cutting fluid, respectively, surface roughness reduced 28.5% and 18.3% in CNT/MoS_2 hybrid nano-cutting fluid at 2% weight concentration. CNT/MoS_2 hybrid nano-cutting fluid at 3% weight concentration decreased cutting temperature by 43.4% and 28%, and tool flank wear by 81.3% and 75%, compared to dry machining and traditional cutting fluid, respectively. Singh et al. [82] turned AISI 304 stainless steel using the MQL technique. The performance of the base fluid (5% Servocut S oil + deionized water), alumina based nanofluid, and hybrid nanofluid (alumina + graphene platelets) were compared. In comparison to the alumina nanofluid and base fluid, the hybrid nanofluid delivered improved machining forces and surface quality. Figure 4.16 shows that the cutting force (Fz), thrust force (Fy), feed force (Fx) and surface roughness (Ra) were reduced by 9.94%, 17.4%, 7.25% and 20.3% respectively under hybrid nanofluid in comparison to alumina nanofluid.

Ngoc et al. [122] studied the hard turning performance of 90CrSi steel (60–62 HRC) in an MQL environment using a hybrid Al_2O_3/MoS_2 fluid, and the results were compared to those obtained using only Al_2O_3 and MoS_2 nanofluids. Hybrid nano-cutting oils in MQL have been proven to improve hard machining performance when compared to mono nanofluids. Because the responses were the lowest and most consistent in MQL, Al_2O_3/MoS_2 hybrid nano-cutting oils improved machining efficiency and yielded superior results to MQL using MoS_2 and Al_2O_3 mono nanofluids. During cooling lubrication conditions, nanoparticle concentration and quadratic interaction had the greatest effects on surface roughness Ra. The quadratic interactions of nanoparticle concentration (NC*NC) and air pressure (P*P) had the greatest influence on rearward force Fp and Fc in MoS_2 nanofluid and Al_2O_3/MoS_2 hybrid nanofluid, whereas in Al_2O_3 nanofluid the interaction effect of air flow rate (NC*Q) and air pressure had a considerable impact. The concentration of MoS_2 nanoparticles has a substantial effect on rearward force Fp. At a tolerable concentration value (approximately 0.5%, the average measured range), air pressure and air flow rate had the least effect, which also stabilized the reactions. However, the back force Fp can be used to evaluate cooling and lubrication strategies.

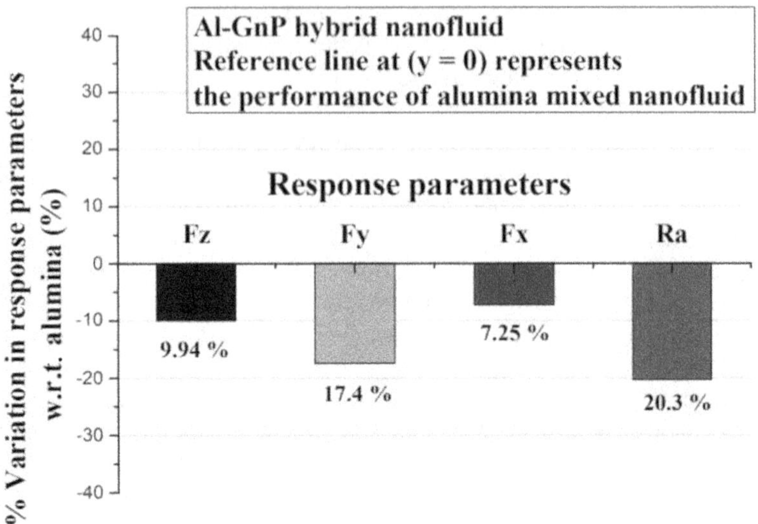

FIGURE 4.16 The influence of alumina–graphene platelets hybrid nanofluid on response characteristics (Fz – cutting force, Fy – thrust force, Fx – feed force and Ra – surface roughness) in comparison to alumina nanofluid. (From Ref. [82].)

Khan et al. [123] compared the performance of MQL, NFMQL and hybrid NFMQL (HNFMQL) conditions during hard turning operations and analyzed surface roughness, tool wear, cutting power and tool life of the hardened steel after the experiment. Al_2O_3 and Al–GnP nanofluid and hybrid nanofluids with different concentrations are used in this experiment. From the results analysis it was observed that the utilization of HNFMQL in machining operations resulted in a notable enhancement of the surface quality of the workpiece. From Figure 4.17, it can be clearly observed that, at all cutting speeds, Al–GnP HNFMQL provided better outcomes regarding surface improvement than both the MQL only and NFMQL environments. The augmentation of cutting velocity resulted in an increase in both cutting efficacy and machine tool potency across all modes of lubrication. The utilization of HNFMQL in machining resulted in lower power consumption across all cutting speeds. The adoption of HNFMQL in hard turning resulted in lower specific energy consumption and specific cutting energy values compared to MQL and NFMQL across all cutting conditions. The specific energy consumption decreased by 50% when the volume % concentration increased from 0.2 to 0.7, while maintaining a constant cutting speed. 25% of the electrical energy utilized by the machine tool was attributed to the cutting process. The implementation of HNFMQL resulted in an enhancement of tool longevity and optimization of material removal volume with respect to tool life analysis.

The use of hybrid nanofluids in hard machining also poses some challenges. For example, the stability and dispersion of nanoparticles in the base fluid can affect the overall performance of the coolant. Furthermore, the presence of nanoparticles can cause issues such as clogging of the coolant system and increased tool wear due to the

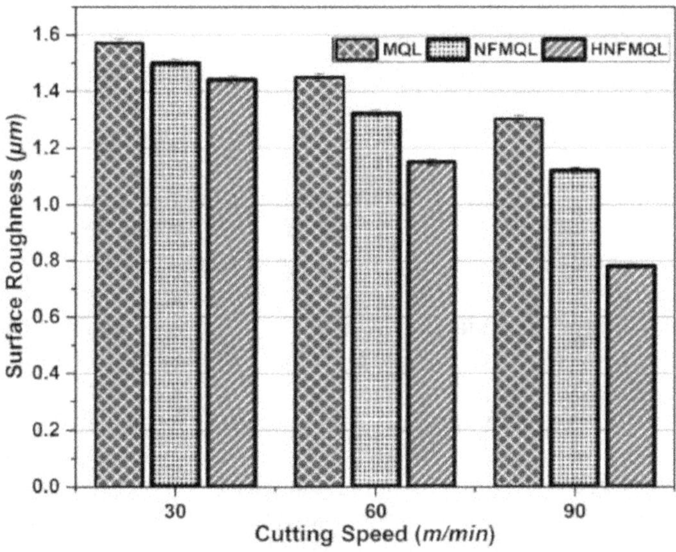

FIGURE 4.17 Comparison graph of surface roughness under MQL, NFMQL and HNFMQL with different cutting speeds. (From Ref. [123].)

abrasive nature of some nanoparticles. Overall, the application of hybrid nanofluids in hard machining is an active area of research, and more studies are needed to fully understand their potential benefits and limitations.

4.5 SUSTAINABILITY ASSESSMENT

Sustainability assessment in nanofluid assisted hard machining involves evaluating the environmental and economic impacts of the process. The use of nanofluids in hard turning has been shown to reduce tool wear, increase material removal rate and improve surface finish. However, the sustainability of the process needs to be evaluated to ensure that it does not have adverse effects on the environment and the economy. The use of nanofluids in hard turning can improve the sustainability of the process in several ways.

Firstly, nanofluids can reduce the amount of cutting fluid required in the hard turning process, resulting in less waste generation and lower environmental impact. Traditional cutting fluids contain harmful chemicals and metals that can be hazardous to the environment if not disposed of properly. The use of nanofluids, on the other hand, can reduce the amount of waste generated and help minimize the environmental impact of the process [124].

Secondly, the use of nanofluids can lead to reduced tool wear and increased material removal rate, which can improve the efficiency of the process. This can lead to reduced energy consumption and lower greenhouse gas emissions, contributing to the overall sustainability of the process. The use of nanofluids can also improve the quality of the machined parts, resulting in less material waste and better product performance. This can help reduce the overall environmental impact of the process by minimizing the need for additional manufacturing or repairs [125].

Although the literature reveals that nanofluids have significant potential to replace traditional cooling strategies for hard machining processes, there are several environmental issues. Therefore, the environmental impact of the nanofluids themselves needs to be evaluated.

To ensure the sustainability of hard machining using nanofluid, it is essential to consider the entire life cycle of the process, from the production and disposal of the nanofluids to the end-of-life disposal of the machined parts. An environmental impact assessment should be conducted to identify potential environmental impacts and develop strategies to minimize them. Additionally, efforts should be made to reduce energy consumption and optimize the use of resources to ensure the overall sustainability of the process [126].

Several researchers have reported hard machining sustainability assessments using Pugh matrix and life cycle assessment decision making tools. Dash et al. [127] used the Pugh matrix tool to compare the sustainability of dry and NFMQL conditions while taking into account various factors, including operator safety, surface finishing quality, cutting temperature, noise level, environmental impact, cutting fluid cost, coolant recycling and disposal, and cleaning factors. According to their relative relevance, each of the aforementioned factors that are taken into consideration to be sustainable is given a particular weightage in the form of a numerical value that ranges from 2 to –2. The scoring of each impact factor is done using the worst-to-best standard. It implies that "2" is given for the majority of superior outcomes while "–2" is given for less desirable outcomes. Similar to this, "–1" and "1" scores have been assigned for poorer and more favourable outcomes. Finally, the combined score for the dry and NFMQL conditions was determined. After computation, the NFMQL's net total score is traced to be 9, while for the dry condition, it was found to be 4. Therefore, from the perspective of sustainability, it can be said that machining under NFMQL cooling-lubrication conditions is both economically and socio-technologically advantageous. In another study, Padhan et al. [128] applied a Pugh matrix to measure sustainability in machining hardened AISI D3 steel under dry and graphene based nanofluid cutting scenarios. Under the graphene NFMQL condition, lower noise levels, controlled power consumption, improved surface finish, improved worker safety, reduced environmental impact, excellent part cleaning and almost no coolant cost and coolant disposal were found in contrast to dry hard turning. Therefore, it can be said that the application of the NFMQL cooling-lubrication technique in hard machining applications is environmentally and technologically advantageous. Dash et al. [129] also applied the same Pugh matrix to measure sustainability. It was stated that the effective cooling-lubrication strategy, safer and cleaner manufacturing, environmental friendliness and aid in increasing sustainability were all benefits of using NFMQL for machining. In summary, the suggested NFMQL turning technique is a reliable approach supported by statistical analysis for significant industrial applications, particularly in the mould and die production industries.

Abbas et al. [130] used MQL to examine the sustainability of the hard turning process in dry, flood and alumina based nanofluid settings. They took environmental effects, cutting expenses, waste management, safety and health into consideration. In this study, the environmental performance measure is taken as carbon dioxide emissions (COE). Toxic chemical exposure (TCE), which is a lower-the-better indication, and high-speed surface exposure (HSSE), which is a higher-the-better indicator, are the chosen sustainable indicators for human health and operation safety. HSSE values are assigned as "1", "2" and "3" at 100 m/min, 125 m/min and 150 m/min cutting speeds, respectively. In this study, all sustainable indicators and the observed machining outputs (surface roughness and power consumption) were given identical weighting factors. When employing MQL nanofluid, COE values were set to "3" instead of "1" for flood assisted tests and "2" for dry cutting testing. Additionally, when completing the sustainability analysis of the current cutting trials, the machining expenses for each run were taken into account. The quantity of coolant utilized in terms of waste management has been specified as "1" for dry cutting tests, "2" for MQL nanofluid testing, and "3" when employing flood coolant by taking the lower-the-better principle. The highest sustainability performance is provided by MQL nanofluid, which has a total weighted sustainability index (TWSI) of 0.7. Dry cutting comes in second with a TWSI of 0.52, while employing flood coolant results in the worst sustainability performance with a TWSI of 0.4. Khan et al. [123] carried out energy based cost integrated modelling and sustainability evaluation of AISI 52100 steel turning aided by Al–GnP hybrid nanofluids. Three distinct cutting fluids (base fluid, base fluid with Al_2O_3 nanoparticles, and mixed nanofluid with Al–GnP nanoparticles) were employed in hard turning with a set fluid flow rate of 300 ml/h and compressed air at 6 bars. A comprehensive comparison of 15 measures for sustainability measurement was carried out and is disclosed in Figure 4.18. Because the

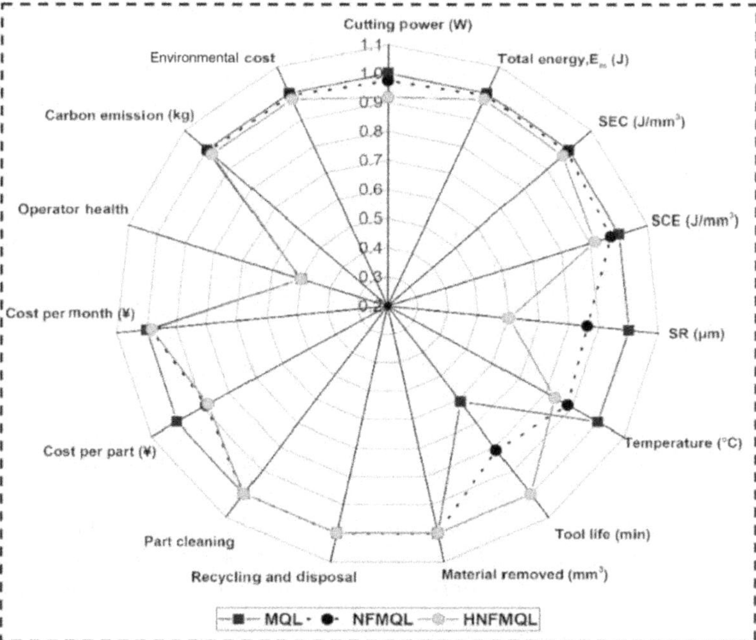

FIGURE 4.18 Overall performance comparison between MQL, NFMQL and HNFMQL with fixed cutting conditions (speed = 90 m/min; feed = 0.1 mm/rev; depth of cut = 0.5 mm and nanoparticle concentration = 0.7). (From Ref. [123].)

MQL, NFMQL and HNFMQL approaches use the least amount of oil and water, there is no need for part cleaning, reuse, recycling or disposal. Cutting tool life, material elimination and user health have to be greater-than-average, but all other measures must be lower. The use of hybrid nanoparticles not only reduces production costs and CO_2 emissions but additionally enhances surface quality dramatically. Overall, the hybrid NFMQL assisted procedure outperformed the MQL and NFMQL techniques.

4.6 INFERENCES

The use of nanofluids as cutting fluids in hard turning has been the subject of intense research in recent years. This chapter offered information on the stability, characterization and production of nanofluids. Additionally, it reported on how well different nanofluids (single and hybrid types) performed in hard machining. In-depth discussion was given on the idea of evaluating sustainability in hard machining using nanofluid cutting fluid. The following major conclusions are drawn:

- The production of nano-cutting fluids has the potential to significantly enhance the efficiency of hard machining operations. The synthesis of nano-cutting fluids for hard turning involves the dispersion of nanoparticles in a base fluid such as water, oil or emulsions. The selection of the base fluid and the nanoparticles can have a significant impact on the properties and performance of the nanofluid.
- Two methods, namely one-step and two-step, are widely implemented to synthesize nanofluids. Two-step methods are easier and make it simple to fabricate nanofluids. Therefore, they are more commonly utilized for the synthesis of nano-cutting fluids for hard machining applications.
- Several properties like thermal conductivity, viscosity, wettability, tribological parameters (wear and friction coefficient) and density are required to describe the characteristics of synthesized nano-cutting fluids.
- The stability of a synthesized nanofluid was traced to be the main issue faced by researchers, especially when it was synthesized using a two-step methodology. Several tests like zeta potential, UV–Vis spectrophotometry, visual sedimentation, pH, 3-omega, centrifugation and electron microscopy (SEM/TEM) are utilized to check the stability of nanofluids. Out of these methods, zeta potential, UV–Vis spectrophotometry, visual sedimentation and electron microscopy are commonly implemented when a nanofluid is used for hard machining applications.
- The most common materials utilized in the synthesis of nanofluids for coolant application in hard machining are metal oxides (copper oxide, aluminium oxide, silicon oxide, zirconium oxide, graphene oxide, zinc oxide, iron oxide and titanium oxide), carbon nanotubes, and molybdenum disulphide. The application of magnesium oxide based nanofluid in hard machining has not been investigated yet, although it has good thermal conductivity and low density.
- The use of nano-cutting fluids has improved machining performance in a promising way. In comparison to conventional cutting fluids, these fluids have shown superior cooling and lubrication capabilities, which reduce tool wear, enhance surface finish and increase material removal rates.

- Nano-cutting fluids can also lead to environmental benefits due to their biodegradability and lower toxicity compared to conventional cutting fluids.
- In hard turning processes, hybrid nanofluids outperformed mono nanofluids in terms of machinability characteristics. However, further research into the synthesis and application of various hybrid nanofluids in hard turning is possible in the future.
- The use of nanofluids in hard turning operations has been found to improve sustainability. Nanofluids successfully reduced cutting force and cutting temperature, resulting in extended tool life, improved surface finishing of machined components, reduced energy consumption, reduced machining expenses, reduced waste and reduced negative environmental effects.
- Very limited research has reported sustainability analysis of nano-cutting fluid assisted hard machining. A Pugh matrix coupled with a Kiviat radar diagram is preferably implemented to analyze the sustainability of this process. This method's accuracy is determined by assigning the proper weight to each factor related to the process.
- The life cycle assessment (LCA) tool is becoming a popular industrial tool to analyze sustainability for the machining process. In hard turning research, LCA was used in just a few studies; hence, it is suggested for future research in hard turning.
- To ensure the efficiency and sustainability of nano-cutting fluids, more study is needed to optimize their composition and use. Furthermore, the economic feasibility of using these fluids must be assessed in order to determine their applicability in manufacturing facilities.

Overall, the utilization of nano-cutting fluids has the potential to revolutionize the machining sector, resulting in more efficient and environmentally friendly machining operations.

REFERENCES

[1] S. Chinchanikar, S.K. Choudhury, Machining of hardened steel—experimental investigations, performance modeling and cooling techniques: A review. *International Journal of Machine Tool Manufacturing,* 89, 95–109, 2015.

[2] R. Mallick, R. Kumar, A. Panda, A. K. Sahoo, Current status of hard turning in manufacturing: Aspects of Cooling Strategy and Sustainability. *Lubricants*, 11, 108, 2023.

[3] B. D. Ma, A. Banerjee, A review of nanofluid synthesis. In: Balasubramanian, G. (eds.), *Advances in Nanomaterials*, pp. 135–176. Springer, Cham, 2018.

[4] V. Fuskele, R. M. Sarviya, Recent developments in nanoparticles synthesis, preparation and stability of nanofluids, materials today. *Proceedings,* 4, 4049–4060, 2017.

[5] V. Sridhara, L. N. Satapathy, Al2O3-based nanofluids: A review, *Nanoscale Research Letters*, 6, 456, 2011.

[6] Y. Wei, X. Huaqing. A review on nanofluids: Preparation, stability mechanisms, and applications, *Journal of Nanomaterials*, Article ID 435873, 2012.

[7] S. U. S. Choi, J. A. Eastman, Enhancing thermal conductivity of fluids with nanoparticles. *ASME-Publications-Fed 231*, 99–106, 1995.

[8] W. Yu, D. M. France, J. L. Routbort, S. U. S. Choi, Review and comparison of nanofluid thermal conductivity and heat transfer enhancements. Heat Transfer Engineering, 29(5), 432–60, 2008.
[9] A. K Sharma, A. K. Tiwari, A. R. Dixit, Rheological behaviour of nanofluids: A review. *Renewable and Sustainable Energy Reviews*, 53, 779–791, 2016.
[10] R. Alimorad, H. Ahmadi, S.S. Mohtasebi M. Pourkhalil, Thermal and rheological properties of oil-based nanofluids from different carbon nanostructures. International Communications in Heat and Mass Transfer, 48, 78–82, 2013.
[11] A. Nasir, M. S. Niasar, A. M. Rashidi, R. Khodafarin, Effect of CNT structures on thermal conductivity and stability of nanofluid. *International Journal of Heat and Mass Transfer*, 55, 1529–35, 2012.
[12] M. Pavía, K. Alajami, P. Estellé, A. Desforges, B. Vigolo, A critical review on thermal conductivity enhancement of graphene-based nanoflfluids, *Advances in Colloid and Interface Science*, 294, 102452, 2021.
[13] S. Lee, S. U. Choi, Application of metallic nanoparticle suspensions in advanced cooling systems (No. ANL/ET/CP-90558; CONF-961105-20). Argonne National Lab. (ANL), Argonne, IL (United States), 1996.
[14] Z. S. Hu, R. Lai, F. Lou, L. G. Wang, Z. L. Chen, G. X. Chen, J. X. Dong, Preparation and tribological properties of nanometer magnesium borate as lubricating oil additive. *Wear*, 252(5–6), 370–374, 2002.
[15] L. Rapoport, V. Leshchinsky, M. Lvovsky, O. Nepomnyashchy, Y. Volovik, R. Tenne, Mechanism of friction of fullerenes. *Industrial Lubrication and Tribology*, 54(4), 171–176, 2002.
[16] G. Liu, X. Li, B. Qin, D. Xing, Y. Guo, Fan, R. Investigation of the mending effect and mechanism of copper nano-particles on a tribologically stressed surface. *Tribology Letters*, 17(4), 961–966, 2004.
[17] X. Tao, Z. Jiazheng, X. Kang, The ball-bearing effect of diamond nanoparticles as an oil additive. Journal of Physics D: Applied Physics, 29(11), 2932–2937, 1996.
[18] K. Lee, Y. Hwang, S. Cheong, Y. Choi, L. Kwon, J. Lee, S. H. Kim, Understanding the role of nanoparticles in nano-oil lubrication. *Tribology Letters*, 35, 127–131, 2009.
[19] D. X. Peng, Y. Kang, R. M. Hwang, S. S. Shyr, Y. P. Chang, Tribological properties of diamond and SiO2 nanoparticles added in paraffin. *Tribology International*, 42(6), 911–917, 2009.
[20] A. H. Elsheikh, M. Abd Elaziz, S. R. Das, T. Muthuramalingam, S. Lu, A new optimized predictive model based on political optimizer for eco-friendly MQL-turning of AISI 4340 alloy with nano-lubricants. *Journal of Manufacturing, Process*, 67, 562–578, 2021.
[21] Y. E. Ç. Vakkas, Experimental comparison of the performance of nanofluids, cryogenic and hybrid cooling in turning of Inconel. Tribology International, 625, 137, 366–378, 2019.
[22] A. R Amin, A. Ali, H. M. Ali, Application of nanofluids for machining processes: A comprehensive review. *Nanomaterials,* 12, 4214, 2022.
[23] Deepika, Nanotechnology implications for high performance lubricants. *SN Applied Sciences*, *2(6)*, Article number: 1128, 2020.
[24] Z. A. A. Ali, A. M. Takhakh, M. Al-Waily, A review of use of nanoparticle additives in lubricants to improve its tribological properties. *Materials Today Proceedings*, 52(3), 1442–1450, 2022.
[25] Y. Li, J. Zhou, S. Tung, E. Schneider, S. Xi, A review on development of nanofluid preparation and characterization. *Powder Technology*, 196(2), 89–101, 2009.

[26] N. Ali, J. A. Teixeira, A. Addali, A review on nanofluids: Fabrication, stability, and thermophysical properties. *Journal of Nanomaterials*, 133, 2018.

[27] P. V. Kazakevich, A. V. Simakin, V. V. Voronov, G. A. Shafeev, Laser induced synthesis of nanoparticles in liquids. *Applied Surface Science*, 252(13), 4373–4380, 2006.

[28] T. X. Phuoc, Y. Soong, M. K. Chyu, Synthesis of Ag-deionized water nanofluids using multi-beam laser ablation in liquids. *Optics and Lasers in Engineering*, 45(12), 1099–1106, 2007.

[29] A. R. Sadrolhosseini, S. A. Rashid, A. Zakaria, K. Shameli, Green fabrication of copper nanoparticles dispersed in walnut oil using laser ablation technique. *Journal of Nanomaterials*, 2016, Article ID 8069685, 2016.

[30] R. Torres-Mendieta, R. Mondragón, V. Puerto-Belda, O Mendoza-Yero, J. Lancis, J. E. Juliá, G. Mínguez-Vega, Characterization of tin/ethylene glycol solar nanofluids synthesized by femtosecond laser radiation. *ChemPhysChem*, 18(9), 1055–1060, 2017.

[31] S. Y. Chun, I. C. Bang, Y. J Choo, C. H. Song, Heat transfer characteristics of Si and SiC nanofluids during a rapid quenching and nanoparticles deposition effects. *International Journal of Heat and Mass Transfer*, 54(5–6), 1217–1223, 2011.

[32] S. W. Lee, S. D. Park, I. C. Bang, Critical heat flux for CuO nanofluid fabricated by pulsed laser ablation differentiating deposition characteristics. *International Journal of Heat and Mass Transfer*, 55, (23–24), 6908–6915, 2012.

[33] V .Piriyawong, V. Thongpool, P. Asanithi, P. Limsuwan, Preparation and characterization of alumina nanoparticles in deionized water using laser ablation technique. *Journal of Nanomaterials*, 1–6, 2012.

[34] V. Thongpool, P. Asanithi, P. Limsuwan, Synthesis of carbon particles using laser ablation in ethanol. *Procedia Enggineering,* 32, 1054–1060, 2012.

[35] M. Y. Choi, D. S. Kim, D. S. Hong, J. H. Kim, Y. T. Kim, Ultrastable aqueous graphite nanofluids prepared by single-step liquid-phase pulsed laser ablation (LP-PLA). *Chemistry Letters*, 40(7), 768–769, 2011.

[36] H. Zeng, X. W. Du, S. C. Singh, S. A Kulinich, S. Yang, J. He, W. Cai, Nanomaterials via laser ablation/irradiation in liquid: A review. *Advanced Functional Materials*, 22(7), 1333–1353, 2012.

[37] H. T. Zhu, Y. S. Lin, Y. S. Yin, A novel one-step chemical method for preparation of copper nanofluids. *Journal of Colloid and Interface Science*, 277(1), 100–103, 2004.

[38] N. Nikkam, M. Ghanbarpour, M Saleemi, E. B. Haghighi, R. Khodabandeh, M. Muhammed, M. S. Toprak, Experimental investigation on thermo-physical properties of copper/diethylene glycol nanofluids fabricated via microwave-assisted route. *Applied Thermal Engineering*, 65(1–2), 158–165, 2014.

[39] A. K. Singh, V. S. Raykar, Microwave synthesis of silver nanofluids with polyvinylpyrrolidone (PVP) and their transport properties. Colloid and Polymer Science, 286 (14–15), 1667–1673, 2008.

[40] R. Jalal, E. K. Goharshadi, M. Abareshi, M. Moosavi, A. Yousefi, P. Nancarrow, ZnO nanofluids: Green synthesis, characterization, and antibacterial activity. *Materials Chemistry and Physics*, 121(1–2), 198–201, 2010.

[41] H. Akoh, Y. Tsukasaki, S. Yatsuya, A. Tasaki, Magnetic properties of ferromagnetic ultrafine particles prepared by vacuum evaporation on running oil substrate. *Journal of Crystal Growth*, 45, 495–500, 1978.

[42] M. Wagener, B. Murty, B. Guenther, Preparation of metal nanosuspensions by high pressure DC-sputtering on running liquids. In: *MRS Proceedings*. Cambridge University Press, Cambridge, 1996.

[43] J. A. Eastman, S. U. S. Choi, S. Li, W. Yu, L. J. Thompson, Anomalously increased effective thermal conductivities of ethylene glycol-based nanofluids containing copper nanoparticles. *Applied Physics Letters*, 78(6), 718–720, 2001.

[44] T. T. Tsung, H. Chang, L. C. Chen, L. L. Han, C. H. Lo, M. K. Liu, Development of pressure control technique of an arc-submerged nanoparticle synthesis system (ASNSS) for copper nanoparticle fabrication. *Materials Transactions*, 44(6), 1138–1142, 2003.

[45] C. H. Lo, T. T. Tsung, L. C. Chen, C. H. Su, H. M. Lin, Fabrication of copper oxide nanofluid using submerged arc nanoparticle synthesis system (SANSS). *Journal of Nanoparticle Research*, 7, 313–320, 2005.

[46] W. Yu, H. Xie, L. Chen, Y. Li, Enhancement of thermal conductivity of kerosene-based Fe3O4 nanofluids prepared via phase-transfer method. *Colloids and Surfaces A*, 355(1–3), 109–113, 2010.

[47] B. K. Park, S. Jeong, D. Kim, J. Moon, S. Lim, J. S. Kim, Synthesis and size control of monodisperse copper nanoparticles by polyol method. *Journal of Colloid and Interface Science*, 311, 417–424, 2007.

[48] H. Shankar, J. W. Rhim, Effect of copper salts and reducing agents on characteristics and antimicrobial activity of copper nanoparticles. *Materials Letters,* 307, 132, 2014.

[49] P. Gurav, S. S. Naik, K. Ansari, S. Srinath, K. A. Kishore, Y. P. Setty, S. Sonawane, Stable colloidal copper nanoparticles for a nanofluid: Production and application. *Colloids and Surfaces A: Physicochemical and Engineering Aspects*, 441, 589–597, 2014.

[50] Y. Zhao, J. J. Zhu, J. M. Hong, N. Bian, H. Y. Chen, Microwave-induced polyol-process synthesis of copper and copper oxide nanocrystals with controllable morphology. *European Journal of Inorganic Chemistry*, 20, 4072–4080, 2004.

[51] H. Kawasaki, Y. Kosaka, Y. Myoujin, Microwave-assisted polyol synthesis of copper nanocrystals without using additional protective agents. *Chemical Communication,* 47, 7740–7742, 2011.

[52] A. Guzmana, J. Arroyoa, L. Verdea, J. Rengifo, Synthesis and characterization of copper nanoparticles/polyvinyl chloride (Cu NPs/PVC) nanocomposites. *Procedia Materials Science*, 9, 298–304, 2015.

[53] Y. L. Hsin, K. C. Hwang, F. R. Chen, J. J. Kai, Production and in-situ metal filling of carbon nanotubes in water. *Advance Materials*, 13(11), 830–833, 2001.

[54] K. Kimoto, Y. Kamiya, M. Nonoyama, R. Uyeda, An electron microscope study on fine metal particles prepared by evaporation in argon gas at low pressure. *Japanese Journal of Applied Physics*, 2(11), 702, 1963.

[55] M. Pike-Biegunski, P. Biegunski, M. Mazur, Colloid, method of obtaining colloid or its derivatives and applications thereof. Google Patents, 2005. WO2005080030A2, Application PCT/PL2005/000012.

[56] G. J. Lee, C. K. Kim, M. K. Lee, C. K. Rhee, S. Kim, C. Kim, Thermal conductivity enhancement of ZnO nanofluid using a one-step physical method. *Thermochimica Acta*, 542, 24–27, 2012.

[57] M. A. Akhavan-Behabadi, M. Shahidi, M. R. Aligoodarz, An experimental study on heat transfer and pressure drop of MWCNT– water nano-fluid inside horizontal coiled wire inserted tube, *International Communications in Heat and Mass Transfer,* 63, 62–72, 2015.

[58] I. Sharmin, M. A. Gafur, N. R. Dhar, Preparation and evaluation of a stable CNT-water based nano-cutting fluid for machining hard-to-cut material. SN Applied Sciences, 2, 626, 2020.

[59] A. Das, S. K. Patel, S. R. Das, Performance comparison of vegetable oil based nanofluids towards machinability improvement in hard turning of HSLA steel using minimum quantity lubrication. *Mechanics & Industry,* 20, 506, 2019.
[60] Ç. V. Yıldırım, Investigation of hard turning performance of eco-friendly cooling strategies: Cryogenic cooling and nanofluid based MQL. *Tribology International*, 144, 106127, 2020.
[61] A. A. Junankar, Yashpal, J.K. Purohit, G.M. Gohane, J. S. Pachbhai, P. M. Gupta, A. R. Sayed, Performance evaluation of Cu nanofluid in bearing steel MQL based turning operation. *Materials Today: Proceedings,* 44(6), 4309–4314, 2021.
[62] S. Mukherjee, P. C. Mishra, P. Chaudhuri, Stability of heat transfer nanofluids – *A review, ChemBioEng Reviews*, 5(5), 312–333, 2018.
[63] A. Das, O. Pradhan, S. K. Patel, S. R. Das, B. B. Biswal, Performance appraisal of various nanofluids during hard machining of AISI 4340 steel, *Journal of Manufacturing Processes*, 46, 248–270, 2019.
[64] D. Wu, H. Zhu, L. Wang, L. Liu, Critical issues in nanofluids preparation, characterization and thermal conductivity. *Current Nanoscience*, 5, 103–112, 2009.
[65] G. Chen, W. Yu, D. Singh, D. Cookson, J. Routbort, Application of SAXS to the study of particle-size-dependent thermal conductivity in silica nanofluids. *Journal of Nanoparticle Research,* 10, 1109–1114, 2008.
[66] A. B. Andhare, R.A. Raju, Properties of dispersion of multiwalled carbon nanotubes as cutting fluid. Tribology Transactions, 59(4), 663–670, 2016.
[67] A. R. I. Ali, B. Salam, A review on nanofluid: preparation, stability, thermophysical properties, heat transfer characteristics and application. *SN Applied Science*, 2, 1636, 2020.
[68] S. Khatai, R. Kumar, A. K. Sahoo, A. Panda, D. Das, Metal-oxide based nanofluid application in turning and grinding processes: A comprehensive review, *Materials Today: Proceedings*, 26(2), 1707–1713, 2020.
[69] H. Xie, W. Yu, W. Chen, MgO nanofluids: higher thermal conductivity and lower viscosity among ethylene glycol-based nanofluids containing oxide nanoparticles. *Journal of Experimental Nanoscience*, 5(5), 463–472, 2010.
[70] T. H. Kim, G. W. Mulholland, M. R. Zachariah, Density measurement of size selected multiwalled carbon nanotubes by mobility-mass characterization. *Carbon*, 47(5) 1297–1302, 2009.
[71] A. J. Ruys, I. G. Crouch, Siliconized silicon carbide. In: Ruys, A. J. (ed.), *Metal-Reinforced Ceramics*, pp. 211–283. Woodhead Publishing, 2021. ISBN 9780081028698. https://doi.org/10.1016/B978-0-08-102869-8.00007-0
[72] A. A. Balandin, S. Ghosh, W. Bao, I. Calizo, D. Teweldebrhan, F. Miao, C. N. Lau, Superior thermal conductivity of single-layer graphene. *Nano Letters*, 8, 902, 2008.
[73] K. Elsaid, M. A. Abdelkareem, H. M. Maghrabie, E. T. Sayed, T. Wilberforce, A. Baroutaji, A. G. Olabi, Thermophysical properties of graphene-based nanofluids. *International Journal of Thermofluids*, 10, 100073, 2021.
[74] G. Fugallo, A. Cepellotti, L. Paulatto, M. Lazzeri, N. Marzari, F. Mauri, Thermal conductivity of graphene and graphite: Collective excitations and mean free paths. *Nano Letters*, 14(11), 6109–6114, 2014.
[75] S. Tadjiki, M. D. Montaño, S. Assemi, A. Barber, J. Ranville, R. Beckett, Measurement of the density of engineered silver nanoparticles using centrifugal FFF-TEM and single particle ICP-MS. *Analytical Chemistry*, 89(11), 6056–6064, 2017.
[76] A. Schavkan, C. Gollwitzer, R. Garcia-Diez, M. Krumrey, C. Minelli, D. Bartczak, S. Cuello-Nuñez, H. Goenaga-Infante, J. Rissler, E. Sjöström, G. B. Baur, K. Vasilatou, A. G. Shard, Number concentration of gold nanoparticles in suspension: SAXS and

spICPMS as traceable methods compared to laboratory methods. *Nanomaterials*, 9(4), 502, 2019.

[77] R. Kumar, A. K. Sahoo, P. C. Mishra, R. K. Das, Influence of Al_2O_3 and TiO_2 nanofluid on hard turning performance. *The International Journal of Advanced Manufacturing Technology*, 106, 2265–2280, 2020.

[78] R. Padmini, P. Vamsi Krishna, G. Krishna Mohana Rao, Effectiveness of vegetable oil based nanofluids as potential cutting fluids in turning AISI 1040 steel. *Tribology International*, 94, 490–501, 2016.

[79] A. Thakur, A. Manna, S. Samir, Multi-response optimization of turning parameters during machining of EN-24 steel with SiC nanofluids based minimum quantity lubrication. *Silicon*, 12, 71–85, 2019.

[80] G. Wang, Y. Li, E. Wang, Q. Huang, S. Wang, H. Li, Experimental study on preparation of nanoparticle-surfactant nanofluids and their effects on coal surface wettability. *International Journal of Mining Science and Technology*, 32(2), 387–397, 2022.

[81] A. Ahmed, I. M. Saaid, A. A. Ahmed, Evaluating the potential of surface-modified silica nanoparticles using internal olefin sulfonate for enhanced oil recovery. *Petroleum Science*, 17, 722–733, 2020.

[82] R. K. Singh, A. K. Sharma, A. R. Dixit, A. K. Tiwari, A. Pramanik, A. Mandal, Performance evaluation of alumina-graphene hybrid nano-cutting fluid in hard turning, *Journal of Cleaner Production*, 162, 830–845, 2017.

[83] S. Khandekar, M.R. Sankar, V. Agnihotri, J. Ramkumar, Nano-cutting fluid for enhancement of metal cutting performance. *Materials and Manufacturing Processes*, 27(9), 963–967, 2012.

[84] H. C. Hamaker, The London—van der Waals attraction between spherical particles. *Physica*, 4, 1058–1072, 1937.

[85] Y. A. Thaher, S. Satoof, A. Kamal, D. Almani, D. Shaban, G. Kassab, H. Surchi, H. Abu-Qtaish, J. Fatouh, S. A. Ajaleh, Chapter 7 – Instrumental analytical techniques for physicochemical characterization of bio-nanomaterials, *Handbook on Nanobiomaterials for Therapeutics and Diagnostic Applications*, Elsevier, 133–150, 2021.

[86] A. Arifuddin, A. A. M. Redhwan, W. H. Azmi, N. N. M. Zawawi, Performance of Al_2O_3/TiO_2 hybrid nano-cutting Fluid in MQL turning operation via RSM approach. *Lubricants,* 10(12), 366, 2022.

[87] G. Gaurav, G. S. Dangayach, M. L. Meena, A. Sharma, Assessment of stability and thermophysical properties of jojoba nanofluid as a metal-cutting fluid: *Experimental and modelling investigation. Lubricants*, 10, 126, 2022.

[88] A. Pal, S. S. Chatha, K. Singh, Performance evaluation of minimum quantity lubrication technique in grinding of AISI 202 stainless steel using nano-MoS2 with vegetable-based cutting fluid. *International Journal of Advance Manufacturing Technology*, 110, 125–137, 2020.

[89] A. Edelbi, R. Kumar, A. K. Sahoo, A. Pandey, Comparative machining performance investigation of dual-nozzle MQL-assisted ZnO and Al2O3 nanofluids in face milling of Ti–3Al–2.5V alloys. *Arabian Journal for Science and Engineering*, 48, 2969–2993, 2023.

[90] Ç. V. Yıldırım, S. Sirin, T. Kıvak, M. Sarıkaya, A comparative study on the tribological behavior of mono&proportional hybrid nanofluids for sustainable turning of AISI 420 hardened steel with cermet tools. *Journal of Manufacturing Processes*, 73, 695–714, 2022.

[91] Y. Fovet, J.-Y. Gal, F. Toumelin-Chemla, Influence of pH and fluoride concentration on titanium passivating layer: stability of titanium dioxide. *Talanta*, 53, 1053–1063, 2001.

[92] H. Wang, M. Sen, Analysis of the 3-omega method for thermal conductivity measurement. *International Journal of Heat and Mass Transfer*, 52(7–8), 2102–2109, 2009.

[93] A. Ditsch, P. E. Laibinis, D. I. C. Wang, T. A. Hatton, Controlled clustering and enhanced stability of polymer-coated magnetic nanoparticles. *Langmuir*, 21, 6006–6018, 2005.

[94] S. S. Khaleduzzaman, M. R. Sohel, R. Saidur, J. Selvaraj, Stability of Al2O3-water nanofluid for electronics cooling system. *Procedia Engineering*, 105, 406–411, 2015.

[95] N. K. Sinha, R. Madarkar, S. Ghosh, P.V. Rao, Application of eco-friendly nanofluids during grinding of Inconel 718 through small quantity lubrication. *Journal of Cleaner Production*, 141, 1359–1375, 2017.

[96] N. A. C. Sidik, S. Samion, J. Ghaderian, M. N. A. W. M. Yazid, Recent progress on the application of nanofluids in minimum quantity lubrication machining: A review. *International Journal of Heat and Mass Transfer*, 108, 79–89, 2017.

[97] A. K. Sharma, A. K. Tiwari, A. R. Dixit, Effects of Minimum Quantity Lubrication (MQL) in machining processes using conventional and nanofluid based cutting fluids: A comprehensive review. *Journal of Cleaner Production*, 127, 1–18, 2016.

[98] A. Das, S. Patel, B. Biswal, N. Sahoo, A. Pradhan, Performance evaluation of various cutting fluids using MQL technique in hard turning of AISI 4340 alloy steel. *Measurement*, 150, 107079, 2020.

[99] P. Sharma, B.S. Sidhu, J. Sharma Investigation of effects of nanofluids on turning of AISI D2 steel using minimum quantity lubrication. *Journal of cleaner Production*, 108, 72–79, 2015.

[100] A. Gupta, R. Kumar, H. Kumar, H. Garg, Comparative performance of pure vegetable oil and Al2O3 based vegetable oil during MQL turning of AISI 4130. *Materials Today Proceeding*, 28, 1662–1666. 2020.

[101] T. M. Duc, T. T. Long, T. Q. Chien, Performance evaluation of MQL parameters using Al2O3 and MoS2 nanofluids in hard turning 90CrSi steel. *Lubricants*, 7, 40, 2019.

[102] P.J. Liew, A. Shaaroni, J. Razak, M. S. Kasim, M.A Sulaiman, Optimization of cutting condition in the turning of AISI D2 steel by using carbon nanofiber nanofluid. *International Journal of Applied Engineering Research*, 12, 2243–2252, 2017.

[103] P. B. Patole, V. V. Kulkarni, Optimization of process parameters based on surface roughness and cutting force in MQL turning of AISI 4340 using nano fluid. *Materials Today: Proceedings,* 5(1), 104–112, 2018.

[104] A. K. Sharma, A. K. Tiwari, A. R. Dixit, Improved machining performance with nanoparticle enriched cutting fluids under minimum quantity lubrication (MQL) technique: A review. *Materials Today: Proceeding*, 2(4–5), 3545–3551, 2015.

[105] A. N. M. Khalil, M. A. M. Ali, A. I. Azmi, Effect of Al2O3 nanolubricant with SDBS on tool wear during turning process of AISI 1050 with minimal quantity lubricant. *Procedia Manufacturing*, 2, 130–134, 2015.

[106] M. Khajehzadeh, J. Moradpour, M. R. Razfar, Influence of nanolubricant particles' size on flank wear in hard turning. *Material Manufacturing Process*, 34(5), 494–501, 2019.

[107] A. Das, S. K. Patel, B. B. Biswal, N. Sahoo, A. Pradhan, Performance evaluation of various cutting fluids using MQL technique in hard turning of AISI 4340 alloy steel. *Measurement*, 150, 107079, 2020.

[108] A. M. M. Ibrahim, M. A. Omer, S. R. Das, W. Li, M. S. Alsoufi, A. Elsheikh, Evaluating the effect of minimum quantity lubrication during hard turning of AISI D3 steel using vegetable oil enriched with nano-additives. *Alexenderia Engineering Journal*, 61(12), 10925–10938, 2022.

[109] N. M. Tuan, T. B. Ngoc, T. L. Thu, Investigation of the effects of nanoparticle concentration and cutting parameters on surface roughness in MQL hard turning using MoS2 nanofluid. *Fluids*, 6(11), 398, 2021.

[110] M. M. S. Prasad, R. R. Srikant, Performance evaluation of nano graphite inclusions in cutting fluids with MQL technique in turning of AISI 1040 steel. *International Journal of Research Engineering and Technology*, 2(11), 381–393, 2013.

[111] P. J. Liew, A. Shaaroni, J. AbdRazak, M. H. Bakar, Comparison between carbon nanofiber (CNF) nanofluid with deionized water on tool life and surface roughness in turning of D2 steel. *Journal of Advanced Research in Fluid Mechanics and Thermal Sciences*, 46(1), 169–74, 2018.

[112] M. Sayuti, A.A. Sarhan, F. Salem, Novel uses of SiO2 nano-lubrication system in hard turning process of hardened steel AISI4140 for less tool wear, surface roughness and oil consumption. *Journal of Cleaner Production,* 67, 265–276, 2014.

[113] U. M. Duc, T. T. Long, P. Q. Dong, Effect of the alumina nanofluid concentration on minimum quantity lubrication hard machining for sustainable production, Proceedings of the Institution of Mechanical Engineers. *Part C: Journal of Mechanical Engineering Science*, 233, 5977–5988, 2019.

[114] K. K Gajrani, P. S. Suvin, S. V. Kailas, R. S. Mamilla. Thermal, rheological, wettability and hard machining performance of MoS2 and CaF2 based minimum quantity hybrid nano-green cutting fluids. *Journal of Materials Processing Technology*, 266, 125–139, 2019.

[115] A. Chavan, V. Sargade, Surface integrity of AISI 52100 Steel during hard turning in different near-dry environments. *Advances in Materials Science and Engineering*, Article ID 4256308, 2020.

[116] D. T. Minh, L. T. The, N. T. Bao, Performance of Al2O3 nanofluids in minimum quantity lubrication in hard milling of 60Si2Mn steel using cemented carbide tools. *Advances in Mechanical Engineering,* 9(7), 1–9, 2017.

[117] T. M. Duc, P. Q. Dong, T. B. Ngoc, Applied research of Nanofluids in MQL to improve hard milling performance of 60Si2Mn steel using carbide tools. *American Journal of Mechanical Engineering*, 5, 228–233, 2017.

[118] U. Alper, D. Furkan, A. Erhan, Applying Minimum Quantity Lubrication (MQL) method on milling of martensitic stainless steel by using nano Mos2 reinforced vegetable cutting fluid. Procedia – Social and Behavioral Sciences, 195, 2742–2747, 2015.

[119] A. K. Sharma, A. K. Tiwari, A. R. Dixit, R. K Singh, Measurement of machining forces and surface roughness in turning of AISI 304 steel using alumina-MWCNT hybrid nanoparticles enriched cutting fluid. *Measurement*, 150, 107078, 2020.

[120] A. K. Sharma, R. K. Singh, A. R. Dixit, A. K. Tiwari, Novel uses of alumina-MoS2 hybrid nanoparticle enriched cutting fluid in hard turning of AISI 304 steel. *Journal of Manufacturing Processes*, 1(30), 467–82, 2017.

[121] S. Gugulothu, V. K. Pasam, Experimental investigation to study the performance of CNT/MoS2 hybrid nanofluid in turning of AISI 1040 steel. *Australian Journal of Mechanical Engineering*, 20(3), 814–24, 2022.

[122] T. B. Ngoc, T. M. Duc, N. M. Tuan, V. L. Hoang, T. T. Long, machinability assessment of hybrid nano-cutting oil for Minimum Quantity Lubrication (MQL) in hard turning of 90CrSi steel. *Lubricants*, 11(2), 54, 2023.

[123] A. M. Khan, M.K. Gupta, H. Hegab, M. Jamil, M. Mia, N. He, Q. Song, Z. Liu, C.I. Pruncu, Energy-based cost integrated modelling and sustainability assessment of Al-GnP hybrid nanofluid assisted turning of AISI52100 steel. *Journal of Cleaner Production*, 257, 120502, 2020.

[124] R. Shah, K. A. Przyborowski, A. Pai, N. Mosleh, Role of nanofluid Minimum Quantity Lubrication (NMQL) in machining application. *Lubricants*, 10, 266, 2022.

[125] E. Usluer, U. Emiroğlu, Y. F. Yapan, G. Kshitij, N. Khanna, M. Sarıkaya, A. Uysal, Investigation on the effect of hybrid nanofluid in MQL condition in orthogonal turning and a sustainability assessment. *Sustainable Materials and Technologies*, 36, e00618, 2023.

[126] N. Khanna, J. Wadhwa, A. Pitroda, P. Shah, J. Schoop, M. Sarıkaya, Life cycle assessment of environmentally friendly initiatives for sustainable machining: A short review of current knowledge and a case study. *Sustainable Materials and Technologies* 32, e00413, 2022.

[127] L. Dash, S. Padhan, A. Das, S. R. Das, Machinability investigation and sustainability assessment in hard turning of AISI D3 steel with coated carbide tool under nanofluid minimum quantity lubrication-cooling condition. *Proceedings of the Institution of Mechanical Engineers, Part C: Journal of Mechanical Engineering Science*, 235(22), 6496–6528, 2021.

[128] S. Padhan, L. Dash, S. K. Behera, Modeling and optimization of power consumption for economic analysis, energy-saving carbon footprint analysis, and sustainability assessment in finish hard turning under graphene nanoparticle–assisted minimum quantity lubrication. *Process Integration and Optimization for Sustainability,* 4, 445–463, 2020.

[129] L. Dash, S. Padhan, S. R. Das, Experimental investigations on surface integrity and chip morphology in hard tuning of AISI D3 steel under sustainable nanofluid-based minimum quantity lubrication. *Journal of Brazilian Society of Mechanical Science and Engineering*, 42, 500, 2020.

[130] A. T. Abbas, M. K. Gupta, M. S. Soliman, M. Mia, H. Hegab, M. Luqman, D. Y. Pimenov, Sustainability assessment associated with surface roughness and power consumption characteristics in nanofluid MQL-assisted turning of AISI 1045 steel. *The International Journal of Advanced Manufacturing Technology*, 105(1–4), 1311–1327, 2019.

5 Computational Methods and Optimization in Sustainable Hard Machining

5.1 INTRODUCTION

Sustainable machining processes are critical nowadays, but they need a complete and trustworthy foundation. The best sustainable machining technique provides a compromise between machining costs, environmental and social impacts, and product quality. These critical elements are essential for determining industrial sustainability. Metal cutting enterprises use environmentally friendly methods to comply with strict environmental standards. Advanced metal cutting producers must boost efficiency while maintaining safe machining conditions. It is noted that any manufacturing process benefits from optimal process cutting conditions [1]. Competition, more environmental regulations, the demand for improved environmental performance in the supply chain, and a decline in skill levels are all placing pressure on metal machining sectors. Even if output falls, machining industries of all sizes can save money and enhance their environmental performance by incorporating sustainability concepts into their operations. Recent industrial issues are the result of virtually high product consumption/production, but real consumption is lower, so a lot of things are held in warehouses [2]. Machining accounts for around 5% of GDP in developed countries. The indirect influence of machining on surface integrity characteristics and product life is significantly greater. Machining will become increasingly crucial as economic factors shorten product life cycles and increase manufacturing flexibility [3]. Sustainable manufacturing necessitates energy-efficient machining, which must be prioritized by any global competition. Energy-efficient machining operations will be improved by well-recognized energy models, quick and effective energy optimization approaches, and good data utilization [4].

Many industrial sectors are interested in the adoption of sustainable machining technologies. Over the past decade, knowledge based optimization methodologies for appropriately calculating cutting parameters have grown, helping with the meeting of sustainability goals. However, a system that fully evaluates ideal outputs and recommends configurations for sustainable machining is essential. A sustainable machining system's methodology must easily adapt to diverse goals. Sustainability is increasingly needed in different industrial operations. The Triple Bottom Line – environmental, social and economic – supports the above requirements. All manufacturing industry levels – product, process and system – must be addressed to meet

DOI: 10.1201/9781003352389-5

sustainability goals [5]. Sustainable machining methods are important for achieving sustainable development goals like good health, decent work and economic growth, and cheap and cleaner energy. Researchers have tried to achieve sustainable machining processes with the best parameters for cutting. But it is important to have an open way of making decisions that leads to the best long-term machining solutions. With this method, decision-makers can easily switch between clusters and scenarios and choose the best way to cut for each case. The method lets people in charge of making decisions choose the best cutting settings for each situation. The results of the study could be put into the machine tool so that the cutting conditions can be changed in real time based on what the decision maker wants [6]. Every aspect of the machining process must be improved for sustainable manufacturing. Increased tool utilization is one of several factors that must be improved so as to achieve a long-lasting machining process [7].

Industrial quality optimization is required for sustainability. Consequently, recommendations for newer renewable technology and labour methods must strike a balance between the ecological and social aspects of industrial production and economic and technological practicality. The field of sustainability research is uncoordinated. Economic and environmental considerations, cooling fluids and power/energy responses are all addressed in sustainable machining. Metal cutting models are complicated by thermo-mechanical interaction, contact/friction and material failure [8]. A performance based machining optimization system for predicting optimal machining conditions in turning processes has been created [9]. Modelling and optimization of severe machining operations is done using computers. To simulate the cutting process components and tune the most important parameters, computational methodologies are applied. Cutting process mathematical modelling aids in understanding the process as well as planning and optimizing machining operations [10].

Sustainable machining must strike a balance between environmental and economic objectives. Multi-objective optimization is performed using Pareto fronts, which are complicated and inefficient when there are more than two objectives. The types are classified into the three major scenarios in Figure 5.1 [11].

Machining is the most important subtractive technique in manufacturing for producing shape features with high dimensional precision and smooth surface quality. It is critical to determine the optimum input parameters for any machining process in order to achieve the best response values and satisfy manufacturers and end users [12]. Process optimization determines the best input parameters for diverse machining processes to maximize energy and resource use and reduce machine operator risk, ensuring a sustainable production environment [13].

The machining processes in a typical manufacturing setup have been found to allocate 40–50% of their total workload to a range of processes including turning, drilling, boring, milling, reaming, deburring, grinding, honing, lapping, broaching, polishing, knurling and finishing. Additionally, the number of lathes present in a workshop accounts for 30–40% of the entire number of machine tools employed in the workshop [14]. The optimization of advanced machining parameters involves the optimization of specific criteria, such as product quality, maximum production rate, minimum production time and minimum production cost. Developing reliable

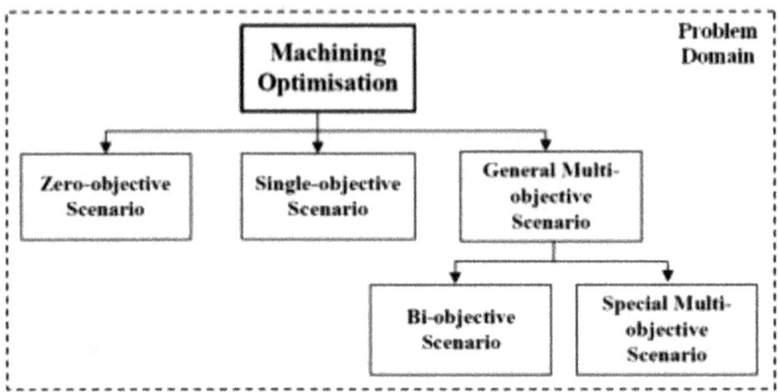

FIGURE 5.1 Classification of problem circumstances. (From Ref. [11].)

models for energy consumption is a crucial aspect in achieving sustainable machining processes. Comprehending and computing the energy expended in machining is a crucial undertaking, as the energy usage during machining constitutes a substantial fraction of the ecological impact of the manufacturing industry [15]. Researchers have developed numerous techniques to address parameter optimization challenges, which can be traditional and non-conventional approaches [16]. The generic classification of various input–output factor relationships and optimization strategies in metal machining operations is depicted in Figure 5.2 (a) and (b).

Optimization of machining operations increases product quality, lowers costs, reduces human error and allows for consistent results. This critical decision-making procedure determines the best option within the constraints. It decreases machining inefficiencies by eliminating the need for machine tool technicians and handbooks to determine machining parameters. As a result, machining unit optimization is more advantageous. The majority of multi-objective optimization problems employ equal weights. Weight assignment methodologies are now being investigated by professionals and researchers working on multi-objective optimization problems. In machining operations, entropy weights were optimized to identify and evaluate continuing and new issues. This strategy benefits researchers/engineers/managers, as well as website/online calculator/software developers working on entropy weight technique advancement [17].

Researchers offer a system for establishing optimal machining settings for milling operations with the primary goal function of optimization of production costs. The circular direction search method, which was specifically built for this purpose and is based on objective function criteria, is used to determine optimal machining settings for each pass. The influence of limitations on the objective function can be analyzed using a graphical image of the objective function and constraints within the generated software [18].

Many studies have assessed the selection of optimal process parameters in sustainable hard machining. For example, Park et al. [19] performed turning experiments for hardened AISI 4140 steel, and these were adjusted in order to reduce cutting energy

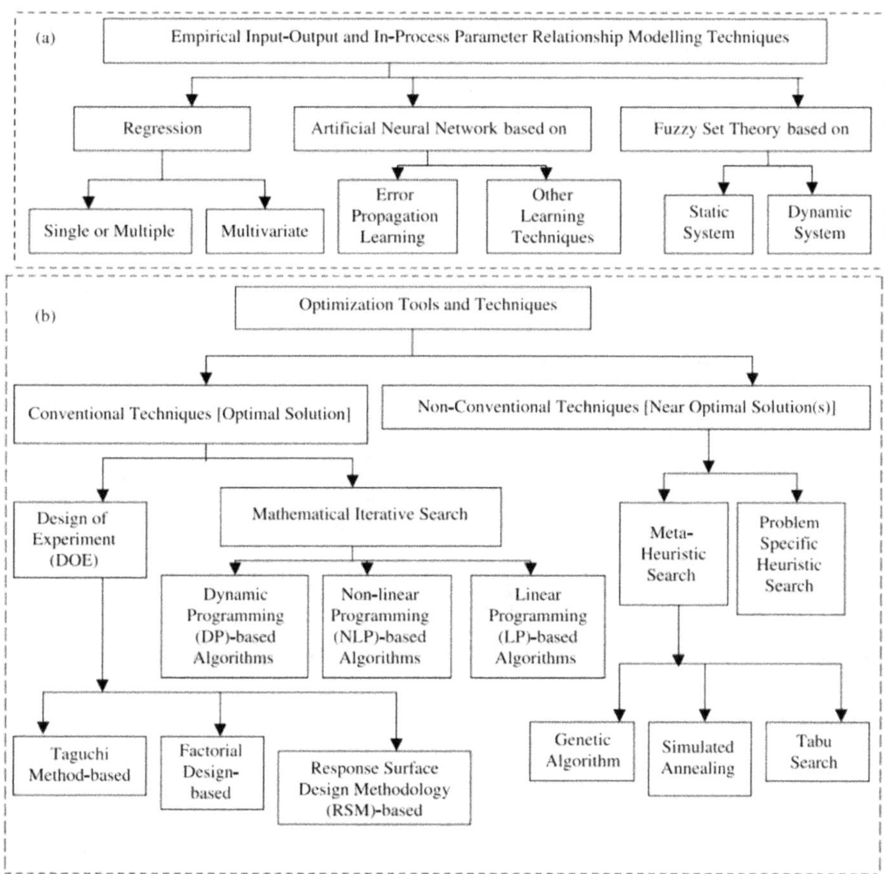

FIGURE 5.2 Classification of (a) modelling and (b) optimization methods in metal cutting practice problems. (From Ref. [16].)

and increase energy efficiency. To demonstrate how machining parameters affect cutting energy and energy efficiency, researchers employed numerical simulations, Box–Behnken design and response surface methodology. The improved system reduces energy by 16% while increasing energy efficiency by 11%.

Das et al. [20] highlighted that many areas want ecologically friendly machining to be optimized. Machining performance depends on input factors, which directly affect process outputs. Selecting input process parameters correctly ensures long-term machining.

Khan et al. [21] recommended a higher machining speed to minimize specific energy consumption. It was further stated that reduced energy usage from machine tool operation saves energy. Figure 5.3 represents the architecture of a performance simulator for sustainable machining processes. The input, process and output are the phases of the architecture.

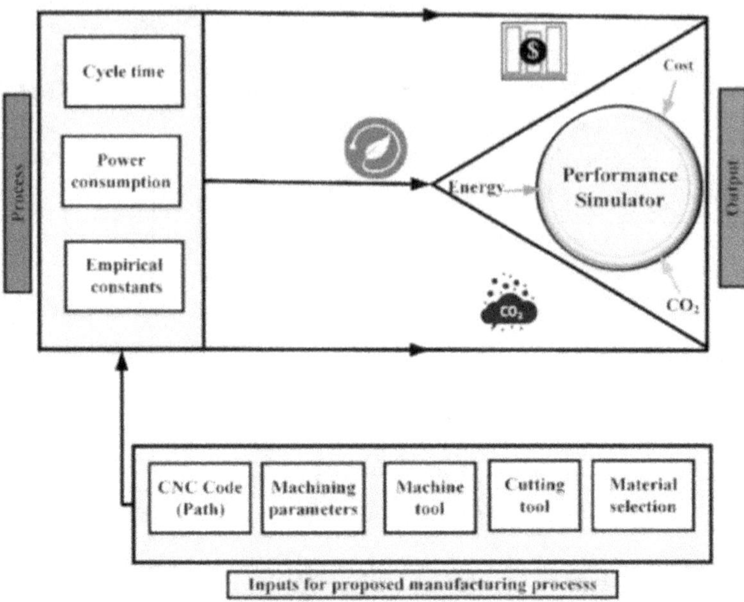

FIGURE 5.3 Architecture of the performance simulator for sustainable machining. (From Ref. [21].)

Manufacturing data are prioritized over simplified physical models and human skills due to the continuously escalating demand and manufacturing complexity processes. Rusinko et al. [22] pointed out that, due to rising energy consumption and carbon emission restrictions, manufacturing firms have conserved machining energy usage. Jayal et al. [23] highlighted that over several product life cycles, sustainability necessitates a comprehensive view of the product, supply chain and production systems. Models, sustainability and the optimization of products, processes and systems all need to improve. The study performed by Ghamdi et al. [24] pointed out how traditional and high-speed machining, cutting speed levels, feed rate and minimum quantity lubrication (MQL) affected various parameters. Cutting tool life, specific cutting energy, process cost, productivity and cutting forces were considered. The two machining modes' effects on manufacturing sustainability were discovered. The study examines machining energy consumption and cutting quality. He et al. [25] studied the energy consumption patterns of machining manufacturing systems and showed how job flow flexibility and variability affect energy consumption. The proposed modelling methodology is applied in Simulink to select flexible task processes to optimize energy consumption. The findings help decision makers assess energy efficiency in machining production systems. Delijaicov et al. [26] pointed out how feed, depth, velocity, MQL and plentiful cooling influenced the machined surface in a heavy part, including microstructural altered layer, residual stresses and roughness statistical parameters. MQL and plentiful cooling caused very compressive circumferential residual strains. Both approaches created microstructure-unaltered surfaces. When measured, the changed surface layer was only 2.35 μm thick. In ideal circumstances, the optimal MQL-changed layer thickness was 74% less than the plentiful-cooling layer. Bharathi et al. [27] highlighted that manufacturing demands require credible models and approaches for estimating the measured performance of cutting processes as this is required for automated machine tool development. In process planning, predicting appropriate machining settings for good surface quality and dimensional accuracy is crucial. Aggarwal et al. [28] designed experimental approaches, notably using response surface methodology (RSM) and Taguchi's technique. The studies were carried out with the use of an L_{27} orthogonal array and a face-centred central composite design. The Taguchi approach and 3D RSM surface plots were used to determine the most important aspects in reducing power usage. According to the findings, the cryogenic environment has the most impact, trailed by cutting speed and depth of cut.

Das et al. [29] performed experiments in dry cutting conditions. AlTiSiN coated carbide tools were coated using a power plasma technique with scalable pulsing, and a dataset was produced of various cutting factors and responses. Furthermore, the machining factors were feed rate, depth of cut and feed rate, and the output responses were surface roughness, cutting force, crater wear length, crater wear breadth and flank wear. The machining data were used to generate surrogate models based on machine learning (ML) that were used to examine, assess and optimize various input machining parameters. Multiple ML techniques, such as polynomial regression, random forest regression, adaptive boosting based regression and gradient boosted trees were utilized to simulate distinct responses in the heavy cutting of AISI D6 steel. Chinchanikar and Choudhury [30] focused on analytical, computational and/or empirical modelling of cutting forces, cutting tool-chip temperature and tool attrition during hardened steel machining under orthogonal/oblique cutting circumstances. The authors presented a complete review of the machining of hard-part steels with coated carbide tools, studies linked to hard-part turning, various cooling and lubrication methods, and efforts made to date to simulate performance. Kalyanakumar et al. [31] evaluated the impact of machining factors on surface roughness and tool vibrations during hard machining, with the process parameters cutting speed 90, 120, 150 m/min; feed rate 0, 1, 0, 3, 0, 5 mm/rev; and depth of cut 0.1, 0.3, 0.5 mm. Using various combinations of cutting settings, the evaluation approach forecasts machined surface quality and tool vibration in the direction of retraction tool wear. Abidi et al. [32] explored how cutting vibration, evaluated surface roughness and tool wear during the dry machining of hard steel with a ceramic tool (mixed) are used to control the parameters online. Further, this analysis is supported by predictive mathematical models. To limit the impact of cutting parameters, for experimental testing, a quality–productivity optimal performance amalgamation of cutting speed, feed rate and depth of cut is chosen from the literature. Abidi et al. [33] compared the hard turning performance of CC650 mixed ceramic, GC1025 coated carbide tool (PVD) and GC4015 coated carbide tool (CVD) materials in a dry cutting environment. The cutting speed was shown to be the most important factor impacting tool life, but feed had the greatest influence on surface roughness. Second order regression models were developed to study the correlations between cutting speed and tool lifetime on one side, and tool lifetime and feed rate on the other.

Ambhore et al. [34] generated a predictive mathematical model by investigating the impact of various machining factors, such as cutting speed, feed and depth of cut, on vibration acceleration and surface roughness. With the aim of designing the experimental runs, the central composite rotatable design approach is used. The empirical findings are then used to build mathematical models through the use of regression analysis. The Design–Expert program was used to complete the work. An artificial neural network (ANN) model was built with the MATLAB program, and the resulting predictions were determined to be within acceptable ranges. A comparison of predicted models and experiments was performed to identify any discrepancies. The results show that the cutting settings have a large and complex influence on the vibration signals. Regression and ANN models have both been found to be effective at predicting vibration-induced effects and surface roughness. The regression coefficient (R^2) obtained is 0.92, demonstrating a strong link between the cutting parameters and the cutting tool vibration. This implies that the created mathematical models are competent at approximating the aforementioned relationship. A conformance test was performed on the resulting correlations, which confirmed their close agreement with the experimental values. A 3.3% minimum deviation was reported. By simplifying difficult machining methods, the ANN model has shown usefulness in duplicating empirical results. According to the findings, the accuracy of predictions made by ANN is superior to that of regression analysis. Neşeli et al. [35] explored the impact of tool geometry on the final surface quality of AISI 1040 steel during the turning process. The study highlighted the impact of tool geometry factors on surface roughness during the turning process. The root mean square technique was used in relation to the average surface roughness (Ra) to construct a prediction model using measured investigational data. In addition, the results indicate that the tool nose radius had the greatest influence on surface roughness. The work proposed a novel intelligence based approach by Pourmostghimi et al. [36] for selecting the most advantageous machining parameters during the hard turning process of AISI D2 material. The study investigates an intelligence-driven technique for increasing tool longevity and increasing material removal rate during AISI D2 finish hard turning procedures. It describes an examination into the link between flank wear and machining parameters during AISI D2 hard finish turning. The work makes use of a genetic equation generated using genetic programming. Gaitonde et al. [37] used second order mathematical models to investigate the effects of depth of cut and cutting time on output characteristics, for example machining force, specific cutting force power, surface roughness and tool wear. Researchers turned AISI D2 cold work (high chromium) tool steel with CC650, CC650WG and GC6050WH ceramic inserts. A full factorial design was used for this study's experimental design. The parametric study findings show that the CC650WG wiper tool outperforms the competition in terms of surface roughness and tool wear. Furthermore, the CC650 conventional insert, on the other hand, is effective at reducing machining force, power and specific cutting force. Aouici et al. [38] explored the impact of cutting speed, feed rate and depth of cut on surface roughness, cutting force, specific cutting force and power during hard turning. The ideal cutting circumstances were determined using RSM and the desirability function technique. According to the findings, using a smaller depth of cut value, a faster cutting speed, and a feed rate of 0.12–0.13 mm/rev during hard turning of AISI D3 hardened steel resulted in reduced cutting forces and improved surface roughness. It is essential to use deeper depths of cut in order to reduce specific cutting force.

Kumar et al. [39] investigated the effects of three different classes of CBN inserts, namely BNX-10 (CBN 1), BN-600 (CBN II) and BNC-300 (CBN III), on cutting forces and surface roughness. The hard turning process was used on die tool steel AISI H13 specimens with varied hardness levels, namely 45 HRC, 50 HRC and 55 HRC. The desirability approach was used to establish the best parameters from a large number of responses. When employing CBN I (BNX-10 grade) inserts, the suggested cutting parameters for obtaining ideal surface roughness, tangential force and thrust force are 180 m/min, 0.08 mm depth of cut, 0.05 mm/rev feed rate and 45 HRC workpiece hardness. Xiao et al. [40] used an orthogonal design and a surrogate model to analyze the impact of spindle speed, feed rate and depth of cut on surface roughness. A dry hard turning process of AISI 1045 steel with a YT5 grade cemented carbide cutting tool was performed. According to the ANOVA and regression model, the feed rate has a greater influence on surface roughness than the other two parameters. Surface roughness and all pairwise combinations of the three variables are revealed through contour and surface plots based on the regression model. The ideal cutting parameters for the specified surface roughness are determined via an optimization process. Altintas et al. [41] pointed out that energy consumption is increasing in lockstep with the global population and degree of optimization, making energy and resource efficiency in manufacturing vital. To enhance energy and resource efficiency, the amount of energy consumed by each industrial operation must first be accurately measured. The pressing need to decrease energy and carbon footprints in manufacturing necessitates an understanding of the lowest environmental footprint production conditions.

Rajemi et al. [42] proposed a fresh structure and approach for improving a manufactured item's energy efficiency. The energy expenditure connected with the turning process for machining a component was optimized and further optimized to determine an economically viable tool life that meets the minimum energy consumption criteria. The study successfully finds crucial elements for optimizing energy use, resulting in lower energy costs and a lower environmental impact. Furthermore, the current study investigates and analyzes the interaction and divergence of economic and environmental elements, as well as the impact of system constraints on determining the most favourable machining settings. Figure 5.4 depicts the ideal tool lifespan that results in the lowest energy usage for two different scenarios. The first situation, known as Case 1, includes the embodied energy of the tool material, but the second scenario, known as Case 2, does not. The study determined that the tool-life duration that results in the lowest energy consumption is 11.4 min, when both the tooling material production and the tool manufacturing process are included.

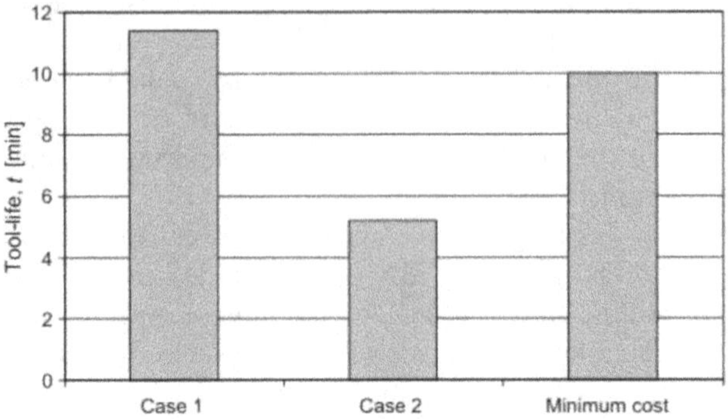

FIGURE 5.4 Optimum tool life with various criteria. (From Ref. [42].)

5.2 COMPUTATIONAL METHODS

Numerous researchers have been drawn to the capabilities of soft computing techniques in dealing with very complicated nonlinear, multidimensional and complex engineering challenges. Soft computing refers to a set of computational methodologies and approaches used to handle specific problem domains. Soft computing methods strive to handle complex problems and provide the best possible answer while accounting for the imprecision and uncertainty inherent in the metal matrix composite machining manufacturing process. Soft computing approaches including the genetic algorithm (GA), response surface methodology (RSM), fuzzy logic (FL), Taguchi method, artificial neural networks (ANN) and particle swarm optimization (PSO) have grown in popularity among researchers. [43]. The organic structure of the human brain was used to construct an ANN. The ANN is one of the most extensively used nonlinear mapping systems in artificial intelligence, capable of tackling a wide range of problems in experimental knowledge, such as modelling, prediction and measurement [44]. The structure of an ANN generally comprises several strata, including the input layer, one or more hidden layers and the output layer. The neurons within layers of processing elements are interconnected by weighted interconnections, which bear a resemblance to the intensity of bioelectricity transmitted among cells in an actual neural network [45]. Öktem [46] studied the simulation and optimization of machining factors for end milling of AISI 1040 steel material under wet conditions with solid carbide tools (TiAlN), with an emphasis on surface roughness. A study was made to evaluate the efficacy of experimental measures and to demonstrate the impact of four cutting factors on surface roughness using multiple regression analysis with analysis of variance. An ANN with the back-propagation learning technique is used to build a surface roughness model. This model was created by implementing a full factorial design of the tests. The optimization technique employs a GA in conjunction with a tested ANN to determine the ideal combinations of machining parameters that result in the desired decreased surface roughness. The application of a GA leads to a reduction of the surface roughness value from 0.67 to 0.59 μm, representing a 12% increase. Gajrani et al. [47] used AISI H13 steel as a workpiece that has undergone hardening, resulting in an average hardness of 56 HRC. The experimental findings indicate that minimum quantity cutting fluids (MQCF) have a significant effect in reducing cutting force, feed force, coefficient of friction and surface roughness. Wang et al. [48] adopted the optimization model considering three objectives: energy, cost and quality. The aforementioned objectives are subject to the influence of three distinct factors, namely cutting depth, feed rate and cutting speed. The machining process is driven by both embodied energy, which includes the energy consumed by cutting tools and cutting fluids, and direct energy, which encompasses startup, cutting and tool change energy, as specified in the model. The expenses associated with machining can be optimized into three distinct groups, namely production running costs, cutting tool costs and cutting fluid costs. Surface roughness is a metric used to assess machining quality. The multi-objective optimization problem is addressed in MATLAB R2011b through the optimization of the NSGA-II approach. Based on the outcomes of the simulation, the optimization of cutting parameters has the potential to enhance energy conservation in machining processes. However, it is important to note that this approach may result in supplementary expenses. Moreover, the impact of optimization on the target of surface roughness is restricted. The subsequent findings serve as an impetus for additional inquiry into the optimization framework that excludes the quality objective. Kulkarni [49] reviewed the applications of the PSO algorithm in the domain of mechanical engineering. Aggarwal et al. [50] used principal component analysis (PCA) to optimize numerous variables (tool service life, surface roughness, cutting force and power consumption) in computer numerical control (CNC) enabled turning of AISI P20 tool steel. The trials were carried out using an L_{27} orthogonal array. The Taguchi approach was used for single response optimization. To correspond to multi-response scenarios, PCA was used. The optimal combination of process components was identified based on the first principal component, which was then analyzed by extracting more than one principal component and amalgamating it into a complete index. Finally, ANOVA was used to determine the most significant CNC turned parameter for multi-response problems.

Kant et al. [51] pointed out that ANN and support vector regression models can estimate machining power usage. An actual machining experiment was done to test these power output predicting methods. ANN outperformed support vector regression in the investigation. Predictive models will help adjust spindle rpm, feed and depth of cut to reduce power usage during milling.

El-Axir et al. [52] proposed an improved experimental model capable of forecasting residual stress profiles in a more thorough manner. The key advantage of this model over previous models is its ability to determine the impact of cutting parameters on maximum residual stress while also pinpointing the specific location and depth of said maximum residual stress. The deflection-etching technique was used to establish the distribution of residual stress in the machined region of the surface. Wang and Feng [53] generated an empirical model capable of consistently forecasting surface roughness in finish turning. The model considers workpiece hardness (material), feed, cutting tool point angle, spindle speed, depth of cut and cutting time. Nonlinear regression analysis and logarithmic data transformation are used to generate the empirical model, which is a fundamental data mining strategy. The aforementioned model's surface roughness values are then validated by additional experiments and compared to those of selected models reported in earlier research. To demonstrate the efficiency of the created model, the authors used metal cutting experiments as well as statistical testing. According to the findings, the proposed model outperforms various current models in terms of error reduction and has acceptable goodness of fit in both model generation and verification. He et al. [54] presented the state of soft computing algorithms for hard turning surface roughness prediction analysis. Further, this focused on data collection, feature selection and a model for predicting surface roughness.

Muthuswamy [55] described the design and production process of a unique wiper milling insert before evaluating its sustainability parameters. Machining power, specific cutting energy, total machining time and tool usage are all factors considered in the evaluation. The study confirms that the recently developed wiper milling insert has the potential to simplify the integration of rough and finish machining operations into a single step while consuming the same amount of electricity. As a result, the quantity of tools required for the machining process may be reduced. Furthermore, the use of the wiper insert can result in a 50% lessening in machining time as well as an important development in surface texture.

Lu et al. [56] used a support vector machine (SVM) to construct prediction models for machining processes. Both the kernel and loss functions are Gaussian radial basis functions with Ɛ-insensitive loss functions where Ɛ stands for the radius of the loss-insensitive hypertube. The artificial bee colony approach (ABC) is used to optimize internal SVM factors so as to improve prediction accuracy and decrease parameter changing time. The experimental results reveal that, with the exception of three DE-SVM indicator values, the selected four evaluation indicator values that reflect prediction accuracy and adjustment time for the ABC-SVM outperform the DE-SVM, GA-SVM and PSO-SVM for AISI 1045 steel. The ABC technique features fewer control factors, and a greater searching ability, than the differential evolution (DE), GA and PSO algorithms for optimizing the internal parameters of an SVM. These findings illuminate how to choose an appropriate SVM optimization technique for industrial processes.

Nguyen et al. [57] emphasized that in order to lower the overall specific energy, carbon emissions and average roughness associated with the aggressively driven rotary turning process, an optimization based on machining conditions should be performed. The processing factors are the tool rotational speed (Tv), depth of cut (a), feed rate (fr) and workpiece speed (Wv). Experiments involving the turning of SKD11 mould material were carried out on a CNC lathe. The regression technique was used to build all-encompassing models for total specific energy, carbon emissions and average roughness. Following that, the entropy algorithm was used to estimate the weightage of individual actively driven rotary turning (ADRT) replies. The non-dominated sorting PSO method is used to determine the best parameters. The study's findings indicate that the best parameters for tool rotational speed, depth of cut, feed rate and workpiece speed are 77 m/min, 0.32 mm, 0.25 mm/rev and 128 m/min, respectively.

Labidi et al. [58] determined the desirability function (DF) approach that should be used to minimize the optimal machining conditions simultaneously (flank wear, arithmetic surface finish and tangential force). The multi-objective optimization of the aforesaid technological parameters using the desirability function found an optimal setting at speed 80 mm/min, feed rate 0.08 mm/rev and time 4 min. Safei et al. [59] machined cold work tool steel X210Cr12 with a carbide tool covered with a triple layer of chemical vapour deposition (CVD) coating, consisting of Al_2O_3, TiC and TiCN. A Taguchi analysis was used to optimize a single objective using the signal-to-noise (S/N) ratio. Furthermore, the study used multi-criteria optimization approaches such as GRA, MOORA, DEAR and WASPAS, with the S/N report serving as the foundation for analysis. The intended goal is to lower Ra, Fz and Pc while simultaneously boosting metal removal rate (MRR). De Souza et al. [60] presented an optimizing technique for analyzing the trade-off between limiting cutting time and the distance between their goals' real mean and total roughness, as well as flank wear variation. Standard roughness optimization approaches in hard turning that ignore tool wear influenced this work. Wear evolution alters roughness, limiting the process's capability to meet specifications and causing cutting tools to be underutilized. The techniques used were the response surface approach, combination array, robust parameter design, multivariate mean square error and normal boundary intersection. The cutting speed, feed rate and depth were used as process parameters during dry finish turning of AISI 52100 hardened steel. Higher cutting times produce the lowest roughness values with the greatest robustness. The proposed technique allows for the investigation of new alternatives to ensure that machined components meet roughness requirements under unchanging cutting time throughout the tool life. The technique must be adaptable to different standards. Torres et al. [61] adopted the new method considering cutting speed, feed rate, depth of cut and stochastic industrial variables such as setup time, batch size, insert changing time, machine and personnel expenses, tool holder price, tool holder life, and insert price. The goal was to model each variable with a probability distribution in order to calculate the total process cost per piece. Adopting a central composite design, the response surface technique was applied to simulate tool service life, average surface roughness and peak-to-valley roughness.

Abbas et al. [62] turned AISI 4340 alloy steel with wiper nose (DCMX 11 T304-WF GC4325) and conventional round nose (DCMT 11 T304-PF GC4325). The multiple objective optimization algorithms multi-objective genetic algorithm, multi-objective Pareto search algorithm and multi-objective emperor penguin colony algorithm were used by the researchers. In terms of computing time, the multi-objective Pareto search algorithm technique outperformed all other insertion strategies, being 30 times faster than the multi-objective genetic algorithm and 20 times faster than the multi-objective emperor penguin colony algorithm. Ahamed et al. [63] evaluated the impact of vibration on ceramic cutting tools under dry conditions on a CNC lathe machine and found the ideal machining situation for the hardened type steel machining process. Furthermore, an integrated fuzzy TOPSIS based Taguchi L_9 optimization model was employed for multiple objective optimization of the hard turning responses. The best multi-objective responses were obtained at a cutting speed of 98 m/min, feed rate of 0.1 mm/rev and depth of cut 0.2 mm.

Motorcu et al. [64] used AISI 8660 steel (50 HRC), a high carbon, chromium–nickel–molybdenum alloy, as the workpiece during machining operation. The optimal surface roughness testing conditions were cutting speed at level 1, feed rate at level 1, depth of cut at level 1, and the nose radius of the tool at level 2. Candioti et al. [65] elucidated the desirability function based on a fundamental principle that considers a product or process to be of unacceptable quality if any of its attributes surpass the desirability limit. When optimizing a process or analytical method, the final solution must be in an optimal region to meet the criteria for each system variable. Raja and Bhaskar [66] developed a method for optimizing machining parameters in an effort to reduce machining time while preserving the appropriate surface roughness. In order to reduce machining time while maintaining the required surface roughness, the best machining parameters were discovered using PSO. Cutting speed, feed, depth of cut and surface roughness are physical restrictions for both the experiment and the theoretical method. It was discovered that PSO is capable of choosing suitable machining variables for turning operations since it was noticed that the machining time and surface roughness based on PSO are almost identical to those attained based on confirmation trials.

Cai et al. [67] generated an analytical model for predicting milling temperature. The suggested model incorporates the cutting factors and can produce the temperature profile at the workpiece surface in less than a minute. It is based on the modified orthogonal cutting model and the boundary layer lubrication results. AISI D2 steel milling experiment results were used to validate the model. Although there were some drawbacks due to the parameter selection needing more investigation and calibration, the suggested model provided reasonable temperature prediction results in a brief calculation period. The suggested design can still serve as a guide for forecasting future temperatures. Wang et al. [68] developed a unique but straightforward generalized optimum estimator based on neural networks for hard turning tool wear prediction. Cutting circumstances and machining time are the inputs to the proposed estimator's fully forward-connected neural network, which produces tool flank wear as the output. To hasten the convergence of learning, the extended Kalman filter technique is used as the network training procedure. Utilizing a destructive optimization approach, network neuron connections are optimized.

5.3 MULTI-RESPONSE OPTIMIZATION TECHNIQUES

Choosing the best parametric combination for a machining process may also be crucial in this regard. Researchers are very interested in using different optimization approaches to choose the best machining parameters. Cica et al. [69] used a grey relational analysis based on the Taguchi method to offer a multi-objective optimization of hard milling settings. For testing coating, cutting speed, feed per tooth and depth of cut, the L_{27} orthogonal array was selected. Surface roughness, MRR and a particular cutting energy were chosen as the highest quality solutions. While PCA was used to assign the proper weight factor to each response variable, grey relational analysis was used to determine the optimal parameter level setting of input parameters for minimizing the specific cutting energy and surface roughness and for maximizing material removal rate. The results of Taguchi based, GRA based optimization revealed that the ideal set of milling parameters is an AlTiN/TiN coating, a cutting speed of 150 m/

min, a feed per tooth of 0.5 m/tooth and depth of cut 1 mm. Iqbal et al. [70] adopted the D-optimal method to quantify the effects of parameters on tool life and surface roughness (in the feed and pick-feed directions), including microstructure of hardened steel, workpiece inclination angle, cutting speed and radial depth of cut. The D-optimal method, which is connected to RSM and used for experiment design, variance analysis and empirical modelling, is a relatively new method.

Rabei et al. [71] performed modelling and surface roughness optimization to enhance the performance of the grinding process of soft steels using MQL technology. The RSM has been used to construct the second order surface roughness estimation model. For the model's creation, three process variables – the depth of cut, cutting speed and feed rate – have been taken into account. Through ANOVA, the model has been examined and validated. Finally, the ideal machining settings produced by the GA were closely in line with the results of the experimental test. Aouici et al. [72] investigated the effects of cutting settings on flank wear (VB) and surface roughness (Ra) when turning hardened AISI H11 (X38CrMoV5-1) steel with a CBN tool. The RSM is used to guide the machining tests. ANOVA is used to examine the combined impacts of three cutting factors, namely cutting speed, feed rate and cutting time, on the two performance outputs (i.e. VB and Ra). The best cutting circumstances for each measured performance level are defined, and a quadratic regression model is used to ascertain the link between the variables and the technological parameters. The findings demonstrate that the cutting duration and, to a lesser extent, the cutting speed have a significant impact on flank wear. Dikshit et al. [73] investigated how TiAlN coated biocompatible and uncoated carbide inserts, and cutting factors such as feed, rotation speed and depth of cut, effect surface roughness in hard-part turning of M2 tool steel (64 HRC). The centre composite design is used in the experimental configuration. Surface roughness testing is performed differently on coated and uncoated inserts. To evaluate a quadratic model, ANOVA is utilized. The Pareto chart ranks cutting criteria in order of importance. A confirmation experiment confirms the cutting parameter optimization of the composite desirability function to limit surface roughness. For TiAlN coated and uncoated carbide inserts, the experimental and anticipated results are 7.93% and 9.36%, respectively. Surface roughness is reduced in TiAlN coated carbide inserts. Rafighi et al. [74] used heat-treated AISI 52100 steel to achieve workpiece hardnesses of 40 HRC and 45 HRC. Dry hard turning experiments were also performed to investigate the effects of workpiece hardness, cutting speed, feed rate and depth of cut on cutting force, surface roughness and sound intensity. For the experiments, a PVD coated carbide turning tool was used. The Taguchi L_{18} design of experiments was used in order to save experimental time, energy and production costs. The results revealed that the feed had the greatest impact on surface roughness, accounting for 96.3% of the total, while also having a 13.8% impact on cutting force. Cutting speed and workpiece hardness were found to contribute 48.3% and 35.1%, respectively, to cutting force. Because higher workpiece hardness necessitated more energy for plastic deformation, cutting force increased with increasing hardness. The depth of cut (53.3%) and cutting speed (40% of the time) had the greatest influence on sound intensity. Finally, the optimum machining parameters were determined using TOPSIS. Rath et al. [75] turned with TiN coated

Al_2O_3+TiO_3+Ti(C,N) ceramic tool inserts to study the machinability of hardened AISI D3 tool steel under dry conditions. The experimental dataset was utilized to create a meaningful regression model for primary process outputs and mean cutting force using RSM. The impacts of chip morphology and parametric interaction on each procedure output were examined. Depth of cut affected cutting force and material removal rate, while tool feed rate affected machined surface quality. At increased feed rates, saw-tooth chips narrowed and developed extensive shear bands. Finally, PSO was used to create an optimal parametric configuration for minimum surface roughness, maximum material removal rate and low cutting force. The ideal parametric settings for surface quality, productivity and power consumption were cutting speed 308 m/min, feed rate 0.08 mm/rev and depth of cut 0.6 mm. Validation showed that this evolutionary PSO method was accurate within 8%. Dureja et al. [76] performed turning of AISI H11 steel with mixed ceramic TiN coated inserts, and this was modelled using the RSM. In the RSM, ANOVA and factor interaction graphs were used to investigate how machining factors such as cutting speed, feed rate, depth of cut and workpiece hardness affect flank wear and surface roughness. The experimental data match well to non-linear quadratic models with optimization of several response factors using the desirability function. Model validation trials predicted response factors to within 5% of the true value. Feed rate, depth of cut and workpiece hardness all have a statistical impact on flank wear, while feed rate and workpiece hardness have an impact on surface roughness. According to Aouici et al. [77], the results show that cutting force components are principally affected by depth of cut and workpiece hardness, whereas surface roughness is statistically affected by feed rate and workpiece hardness.

5.4 HYBRID TECHNIQUES

In turning operations, Sofuoglu et al. [78] studied the surface roughness, cutting pressures and material removal rates of various materials under various cutting conditions. Cutting tests were used to determine the natural and chatter frequencies, stiffness coefficient and damping ratio. Experiments were used to calculate surface roughness, material removal and cutting forces. These experiments propose hybrid multi-criteria decision-making models such as the reference ideal model (RIM), TOPSIS, the multi-criteria optimization and compromise solution (VIKOR), the weighted sum approach (WSA) and ratio based multi-objective optimization (MOORA). Kaladhar et al. [79] adopted the Taguchi technique of experiment design to carry out their experiments. A grey relational grade was revealed in grey relational theory to identify an ideal level of cutting parameters that result in lesser magnitudes of surface roughness, flank wear and tool vibration and a larger magnitude of material removal rate. The combination of machining parameters that improves turning performance was found to be 210 m/min (cutting speed), 0.15 mm/rev (feed rate), 1.0 mm (depth of cut) and 0.4 mm (nose radius). Bhandarkar et al. [80] investigated the machining performance of 100Cr6 bearing steel using current C-type inserts. In experiments, machining factors influenced surface quality, chip reduction coefficient and cutting force. Furthermore, a Taguchi-Satisfaction function distance measure was employed in multiple response optimization. In most cases, WC coated tools outperform ceramics tools. Padhan et al. [81] adopted geometrical characteristics (insert nose radius) and machining parameters (cutting speed, axial feed, depth of cut) to design hard turning trials. To study, predict and improve machining power utilization, central composite design – ANOVA, desirability function analysis and RSM – was utilized.

Cui et al. [82] determined the ideal cutting parameter area in intermittent hard turning, and studies on specific cutting energy, damage equivalent stress and surface roughness were carried out. Based on finite element calculations, micromechanics, damage mechanics and intermittent turning tests, the ideal cutting parameter area was observed.

Frides et al. [83] compared the cutting force components of mixed ceramic [Al_2O_3 (70%) + TiC (30%)] and reinforced ceramic [Al_2O_3 (75%) + SiC (25%)] tools when used to process AISI 4140 steel hardened to 60 HRC. The impacts of cutting speed, feed rate and depth of cut on cutting force components are examined using RSM and ANOVA to model and optimize these technological parameters. According to the findings of this study, utilizing a mixed ceramic insert results in lower cutting force component values than using a reinforced ceramic insert. Zhou et al. [84] proposed a cutting parameter optimization approach for machining operations considering carbon emissions to balance cutting indexes such as cutting time and cutting cost in the component machining process. To solve the problem, an improved algorithm, non-cooperative game theory integrated NSGA-II (NG-NSGA-II), is developed by introducing non-cooperative game theory. Finally, a cylindrical turning is used to demonstrate the logic of the suggested cutting factor optimization procedure. The simulation findings suggest that NG-NSGA-II outperforms NSGA-II. As a result, the suggested method may give the ideal cutting parameters in the component machining process, reducing carbon emissions, cutting time and cutting cost. Bagabar and Yushoff [85] adopted RSM multi-objective optimization and compared it to NSGA-II prior to experimental confirmation tests. RSM's multi-objective optimization saved 9.2% energy and 4.6% machining cost. The second-generation NSGA-II optimization results outperformed RSM by over 70%. A flowchart of multi-optimization using RSM and NSGA-II is shown in Figure 5.5.

Tebassi et al. [86] provided a step-by-step tutorial on modelling, multi-attribute optimization and validation utilizing several approaches. It also comes with a manual

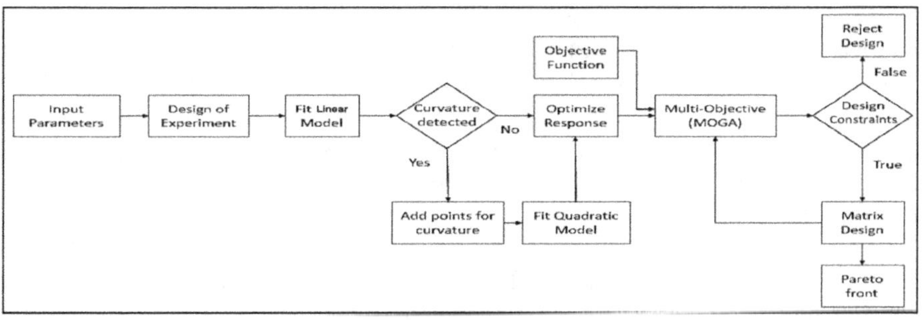

FIGURE 5.5 Flowchart of multi-optimization procedure. (From Ref. [85].).

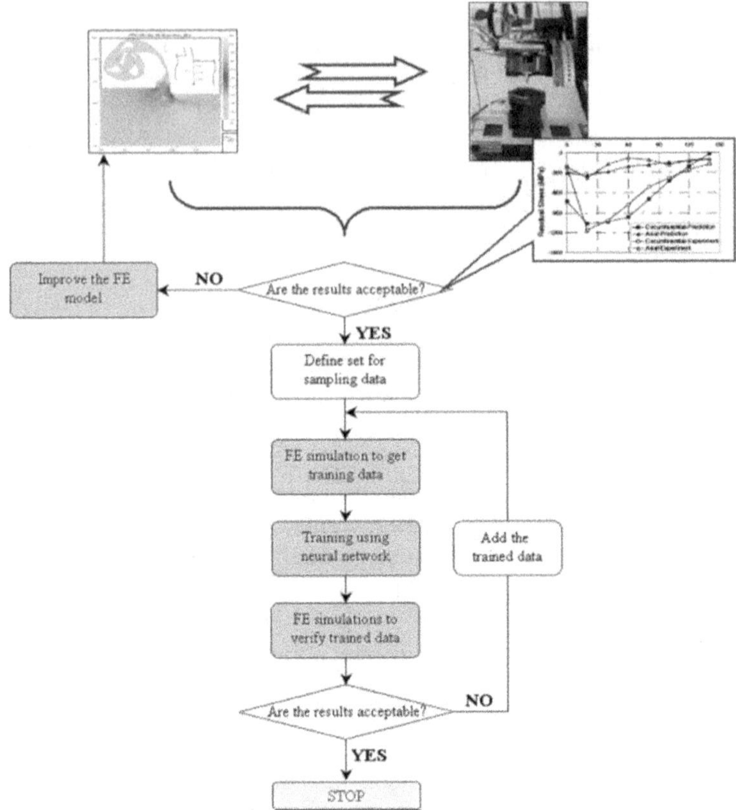

FIGURE 5.6 Proposed ANN training methodology. (From Ref. [87].)

for establishing the best cutting settings over time. Umbrello et al. [87] used ANNs and FEM in a predictive hybrid model for forward and inverse prediction. The former can generate a residual stress profile for a given tool, material and process conditions, whereas the latter can only do so when a residual stress distribution constraint is applied. Three-layer neural networks were developed and validated using numerical data from hard machining of 52100 bearing steel. This shows that hybrid ANN–FEM techniques can be used to establish optimal machining parameters for hardened steels in order to maximize residual stress while minimizing white layer thickness and cutting edge wear. The ANN training method proposed is sketched in Figure 5.6.

Hafiz et al. [88] generated a tool service life model for end milling hardened AISI D2 steel with a coated carbide cutting tool (PVD TiAIN). The hardness of the AISI D2 tool is between 56 HRC and 58 HRC. The key machining parameters chosen as independent variables for this trial were cutting speed, feed and depth of cut. To generate first and second order models, the RSM modelling technique was employed. These models can be used to safely predict the tool life of a machined object produced from AISI D2 steel under the specific machining conditions.

Mia et al. [89] investigated hard turning of 600 BHN steel to study the surface roughness under MQL environment using a variable time-controlled pulse delivering coolant whose flow rate was within 500–1100 ml/h with cutting speed of 66–100 m/min and feed rate of 0.18–0.25 mm/rev respectively. Higher cutting speed with lower feed rate yields minimal surface roughness at MQL pulse of 1 s time gap. The prediction accuracy of the ANN model has been found to be 97.5%. Jiang et al. [90] proposed a model for optimizing machining parameters based on the lowest cutting fluid consumption and cost. The optimization model considers process cost and cutting fluid consumption to be two goals that are affected by depth of cut, feed rate, cutting speed and cutting fluid flow. A hybrid GA approach is utilized to optimize machining parameters with multiple objectives while accounting for cutting fluid consumption and process cost. The hybrid GA method, in contrast to the classic GA approach, has been updated to boost computation performance. Bhattacharya et al. [91] pointed out that dry turning saves energy and machining expenses, making it environmentally friendly. The response values for the considered process are predicted by four prediction models: multivariate regression analysis, FL, ANN and the adaptive neurofuzzy inference system (ANFIS). Five statistical criteria are used to compare their prediction performance: root mean squared percent error, mean absolute percentage error, root mean squared log error, correlation coefficient and root relative squared error. In statistical measure calculations, ANFIS, which mixes FL and ANN, outperforms conventional prediction models. Using diverse machining settings as input variables, the ANFIS model may predict process reactions based on their allowable values. Chudy et al. [92], completed the turning of hardened 41Cr4 alloy steel with variable cutting speed and feed rate utilizing commercial CBN tools, and different cutting forces, cutting energy, specific cutting energy and volumetric machining rates were selected. The machine tool and a unique energy property of severe machining were also considered. This method predicts the relationship between specific cutting energy and volumetric machining removal rate and compares hard machining productivity to other machining techniques. According to Guo et al. [93], finish turning cutting settings are optimized using an energy-efficient and surface roughness based approach. A unique energy model and surface roughness model for a machine tool are used to optimize cutting parameters for a precise surface finish while consuming little energy. The roughness model and energy model are used sequentially to determine the best cutting conditions (Figure 5.7)

Rajabonshi et al. [94] used tungsten coated carbide tools and 48 HRC hardened AISI D2 steel. Cutting speed (v), feed rate (f), depth of cut, surface roughness, flank wear, cutting force and feed force are all process characteristics. Air-cooled turning minimizes flank wear, while dry turning enhances surface roughness, cutting force and feed force. Surface roughness and flank wear are predicted by ANN and RSM models. Bhone et al. [95] adopted the utility idea to optimize the process parameters during the AISI M7 hard-part turning process. Researchers studied the impacts of tool nose radius and cutting speed, feed rate and depth of cut (process parameters) on surface roughness and material removal rate. It was revealed that a cutting speed of 300 m/min, feed rate of 0.15 mm/rev, depth of cut 0.75 mm and a nose radius of 1.2 mm provide the maximum MRR and smallest amount of surface roughness. Lee

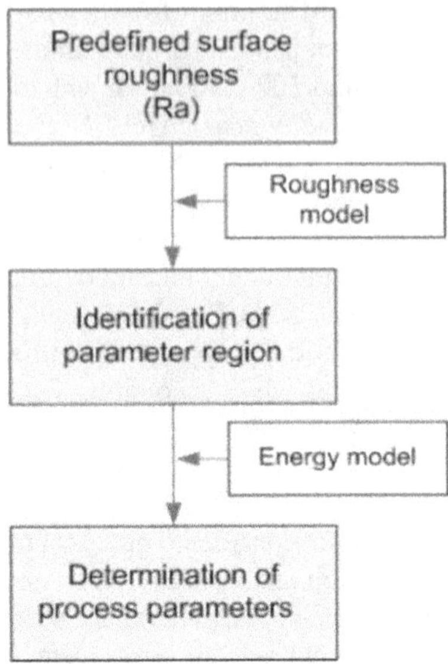

FIGURE 5.7 Diagram of parameter selection in finishing. (From Ref. [93].)

and Badrul [96] used a hybrid model with fuzzy axiomatic design (AD) and crisp techniques to examine sustainability performance. They highlighted that product development and manufacturing organizations can practise environmentally conscious manufacturing as part of sustainable product realization by using the proposed approach to recognize ineffective processes using empirical data.

5.5 INFERENCES

- A great deal of optimization studies and predictive models obtainable during the sustainable turning of hardened steel alloys were done using coated carbide, ceramic and CBN tools as observed.
- When optimal machining parameters are used in machining operations, machined components have better functional qualities and are more productive. Process parameters in the machining process can be optimized to improve process results.
- The suggested knowledge base would be a significant resource for decision-making practitioners and process engineers looking for optimal machining methods for complex hardened engineering materials. This resource would provide in-depth information on important input parameters and responses, as well as potential interactions between them, allowing practitioners to create effective experimental plans depending on the number of input parameters and their respective operating levels.
- Research is expected to yield practical applications for the metal cutting sector within the context of environmentally sensitive and sustainable manufacturing

practices. The goal is to help engineers and machinists determine ideal machining settings that result in higher productivity, lower production costs and lower environmental impact.

- Research results have established that dry machining is a feasible and sustainable approach that can enhance environmental friendliness while being economically viable.
- The innovations outlined in this chapter have the potential to be applied to various production processes, such as the use of cutting fluids, minimum quantity liquid and material consumption. Additional characteristics, such as the amplitude of the cutting force, the temperature at the tool interface and the radial rake angle, may also be considered. The experiments and algorithms given here have the potential to be used to optimize various machining procedures.
- The scope of this study can be expanded to include an examination of the sustainability assessment of additional machining processes. Furthermore, an integrated sustainability index can be developed based on all manufacturing processes being conducted, as the sustainability of the final product is heavily reliant on the enhancement of sustainability in the machining processes involved. The investigation of sustainability disparities among various products produced through advanced machining techniques, such as additive manufacturing (e.g. 3D printing), in comparison to conventional processes, which involve a combination of turning, drilling and milling, presents an intriguing research opportunity.
- This analysis presents an attempt to explain a comparative investigation between Taguchi, Taguchi–grey and Box–Behnken RSM approaches by modelling, analysis and optimization of turning parameters for optimum performance of hardened steel cutting.

REFERENCES

[1] P. Sivaiah, D. Chakradhar, Modeling and optimization of sustainable manufacturing process in machining of 17-4 PH stainless steel. *Measurement*, 134, 142–152, 2019.

[2] J. Kopac, Achievements of sustainable manufacturing by machining. *Journal of Achievements in Materials and Manufacturing Engineering*, 34(2), 180–187, 2009.

[3] A. D. Jayal, F. Badurdeen, O. W. Dillon Jr., I. S. Jawahir, Sustainable manufacturing: Modeling and optimization challenges at the product, process and system levels. *CIRP Journal of Manufacturing Science and Technology*, 2(3), 144–152, 2010.

[4] T. Peng, X. Xu, Energy-efficient machining systems: A critical review. *The International Journal of Advanced Manufacturing Technology*, 72, 1389–1406, 2014.

[5] A. Salem, H. Hegab, S. Rahnamayan, H. A. Kishawy, Multi-objective optimization and innovization-based knowledge discovery of sustainable machining process. *Journal of Manufacturing Systems*, 64, 636–647, 2022.

[6] H. Hegab, A. Salem, H. A. Taha, A decision-making approach for sustainable machining processes using data clustering and multi-objective optimization. *Sustainability*, 14(24), 16886, 2022.

[7] F. Schultheiss, J. Zhou, E. Gröntoft, J. E. Ståhl, Sustainable machining through increasing the cutting tool utilization. *Journal of Cleaner Production*, 59, 298–307, 2013.

[8] M. Vaz, D. R. J. Owen, V. Kalhori, M. Lundblad, L. E. Lindgren, Modelling and simulation of machining processes. *Archives of Computational Methods in Engineering*, 14, 173–204, 2007.

[9] X. Wang, Z. J Da, A. K. Balaji, I. S. Jawahir, Performance-based predictive models and optimization methods for turning operations and applications: Part 3—optimum cutting conditions and selection of cutting tools. *Journal of Manufacturing Processes*, 9(1), 61–74, 2007.

[10] E. Polak, *Computational Methods in Optimization: A Unified Approach*. Mathematics in Science and Engineering series (Vol. 77). Academic Press, New York, 329 p., 1971.

[11] T. Zhang, O. Owodunni, J. Gao, Scenarios in multi-objective optimization of process parameters for sustainable machining. *Procedia CIRP*, 26, 373–378, 2015.

[12] S. Chakraborty, S. Chakraborty, S. A scoping review on the applications of MCDM techniques for parametric optimization of machining processes. *Archives of Computational Methods in Engineering*, 29(6), 4165–4186, 2022.

[13] S. Chakraborty, H. N. Datta, S. Chakraborty, Grey relational analysis-based optimization of machining processes: A comprehensive review. Process Integration and Optimization for Sustainability, 1–31, 2023.

[14] D. D. Trung, D. H. Tien, V. C. C Nguyen, N. T. Nguyen Surface roughness prediction in CNC hole turning of 3X13 steel using support vector machine algorithm. Tribology in Industry 42(4), 597–607, 2020.

[15] T. D. Hoang, N. T. Nguyen, D. Q. Tran, V. T. Nguyen, Cutting forces and surface roughness in face milling of SKD61 hard steel. *Strojniski vestnik-Journal of Mechanical Engineering*, 65(6), 375–385, 2019. https://doi.org/10.5545/sv-jme.2019.6057

[16] I. Mukherjee, P. K. Ray, A review of optimization techniques in metal cutting processes. *Computers & Industrial Engineering*, 50(1–2), 15–34, 2006.

[17] R. Kumar, S. Singh, P. S. Bilga, J. Singh, S. Singh, M. L. Scutaru, C. I. Pruncu, Revealing the benefits of entropy weights method for multi-objective optimization in machining operations: A critical review. *Journal of Materials Research and Technology*, 10, 1471–1492, 2021.

[18] M. C. Cakir, A. Gurarda, Optimization of machining conditions for multi-tool milling operations. *International Journal of Production Research*, 38(15), 3537–3552, 2000.

[19] H. S. Park, T. T. Nguyen, X. P. Dang, Multi-objective optimization of turning process of hardened material for energy efficiency. *International Journal of Precision Engineering and Manufacturing*, 17, 1623–1631, 2016.

[20] P. P. Das, S. Chakraborty, SWARA-CoCoSo method-based parametric optimization of green dry milling processes. *Journal of Engineering and Applied Science*, 69(1), 1–21, 2022.

[21] A. M. Khan, L. Liang, M. Mia, M. K. Gupta, Z. Wei, M. Jamil, H. Ning, Development of process performance simulator (PPS) and parametric optimization for sustainable machining considering carbon emission, cost and energy aspects. *Renewable and Sustainable Energy Reviews*, 139, 110738, 2021.

[22] C. Rusinko, Green manufacturing: An evaluation of environmentally sustainable manufacturing practices and their impact on competitive outcomes. IEEE Transactions on Engineering Management 54(3), 445–454, 2007.

[23] A. D. Jayal, F. Badurdeen, Jr., O. W. Dillon, I. S. Jawahir, Sustainable manufacturing: Modeling and optimization challenges at the product, process and system levels. *CIRP Journal of Manufacturing Science and Technology*, 2(3), 144–152, 2010.

[24] K. A. Al-Ghamdi, A. Iqbal, A sustainability comparison between conventional and high-speed machining. *Journal of Cleaner Production*, 108, 192–206, 2015.

[25] Y. He, B. Liu, X. Zhang, H. Gao, X. Liu, A modeling method of task-oriented energy consumption for machining manufacturing system. *Journal of Cleaner Production*, 23(1), 167–174, 2012.

[26] S. Delijaicov, G. C. Alves, É. C. Bordinassi, E. J. D. Silva, Optimization of machining parameters applied in hard turning of large components for cooling lubrication methods. *Proceedings of the Institution of Mechanical Engineers, Part E: Journal of Process Mechanical Engineering*, 237(2), 580–588, 2023.

[27] S. S. Bharathi, D. Ravindran, A. A. M. Moshi, S. V. Alagarsamy, R. K. Kannan, V. J. Prasath, Machinability study on CNC turning of stainless steel 303 with CVD multi-layer (TiN/Al_2O_3/TiCN) coated carbide insert by using grey-fuzzy logic approach. *Proceedings of the Institution of Mechanical Engineers, Part E: Journal of Process Mechanical Engineering*, 236(5), 1967–1978, 2022.

[28] A. Aggarwal, H. Singh, P. Kumar, M. Singh, Optimizing power consumption for CNC turned parts using response surface methodology and Taguchi's technique—a comparative analysis. *Journal of materials processing technology*, 200(1–3), 373–384, 2008.

[29] A. Das, S. R. Das, J. P. Panda, A. Dey, K. K. Gajrani, N. Somani, N. K. Gupta, Machine learning-based modeling and optimization in hard turning of Aisi d6 steel with advanced AlTiSiN-coated carbide inserts to predict surface roughness and other machining characteristics. *Surface Review and Letters (SRL)*, 29(10), 1–24, 2022.

[30] S. Chinchanikar, S. K. Choudhury, Machining of hardened steel—experimental investigations, performance modeling and cooling techniques: A review. *International Journal of Machine Tools and Manufacture*, 89, 95–109, 2015.

[31] S. Kalyanakumar, A. Elanthiraiyan, S. Sreekanth, S. John, K. E. Afsal, Process optimization of parameters for minimizing vibrations and surface roughness during hard steel by ranking algorithm. *AIP Conference Proceedings,* 2523(1), 020149, 2023.

[32] Y. Abidi, L. Boulanouar, Correlation analysis between tool wear, roughness and cutting vibration in turning of hardened steel. *Engineering Transactions*, 69(4), 403–421, 2021.

[33] Y. Abidi, Machining performance analysis of cutting tool materials in hard turning of bearing steel. *Academic Journal of Manufacturing Engineering*, 19(4), 2021.

[34] N. Ambhore, D. Kamble, S. Chinchanikar, Evaluation of cutting tool vibration and surface roughness in hard turning of AISI 52100 steel: An experimental and ANN approach. *Journal of Vibration Engineering & Technologies*, 8(3), 455–462, 2020.

[35] S. Neşeli, S. Yaldız, E. Türkeş, Optimization of tool geometry parameters for turning operations based on the response surface methodology. *Measurement*, 44(3), 580–587, 2011.

[36] V. Pourmostaghimi, M. Zadshakoyan, Optimization of cutting parameters during hard turning using evolutionary algorithms. *Optimization for Engineering Problems*, 77–99, 2019.

[37] V. N. Gaitonde, S. R. Karnik, L. Figueira, J. P. Davim, Machinability investigations in hard turning of AISI D2 cold work tool steel with conventional and wiper ceramic inserts. *International Journal of Refractory Metals and Hard Materials*, 27(4), 754–763, 2009.

[38] H. Aouici, H. Bouchelaghem, M. A. Yallese, M. Elbah, B. Fnides, Machinability investigation in hard turning of AISI D3 cold work steel with ceramic tool using response surface methodology. *The International Journal of Advanced Manufacturing Technology*, 73, 1775–1788, 2014.

[39] P. Kumar, S. R. Chauhan, C. I. Pruncu, M. K. Gupta, D. Y. Pimenov, M. Mia, H. S. Gill, Influence of different grades of CBN inserts on cutting force and surface

roughness of AISI H13 die tool steel during hard turning operation. *Materials*, 12(1), 177, 2019.

[40] Z. Xiao, X. Liao, Z. Long, M. Li, Effect of cutting parameters on surface roughness using orthogonal array in hard turning of AISI 1045 steel with YT5 tool. *The International Journal of Advanced Manufacturing Technology*, 93, 273–282. 2017.

[41] R. S. Altıntaş, M. Kahya, H. Ö. Ünver, Modelling and optimization of energy consumption for feature based milling. *The International Journal of Advanced Manufacturing Technology*, 86, 3345–3363, 2016.

[42] M. F. Rajemi, P. T. Mativenga, A. Aramcharoen, Sustainable machining: selection of optimum turning conditions based on minimum energy considerations. *Journal of Cleaner Production*, 18(10–11), 1059–1065, 2010.

[43] R. A. Laghari, J. Li, A. A. Laghari, S. Q. Wang, A review on application of soft computing techniques in machining of particle reinforcement metal matrix composites. *Archives of Computational Methods in Engineering*, 27, 1363–1377, 2020.

[44] R. P. Lippman, An introduction to computing with neural nets. *IEEE ASSP* 4, 4–22, 1987. doi:10.1109/MASSP.1987.1165576.

[45] D. Anderson, G. McNeil. *Artificial Neural Network Technology*. Rome Laboratory, New York, 1992.

[46] H. Öktem, An integrated study of surface roughness for modelling and optimization of cutting parameters during end milling operation. *International Journal of Advanced Manufacturing Technology*, 43(9), 852, 2009.

[47] K. K. Gajrani, D. Ram, M. R. Sankar, Biodegradation and hard machining performance comparison of eco-friendly cutting fluid and mineral oil using flood cooling and minimum quantity cutting fluid techniques. *Journal of Cleaner Production*, 165, 1420–1435, 2017.

[48] Q. Wang, F. Liu, X. Wang, Multi-objective optimization of machining parameters considering energy consumption. *The International Journal of Advanced Manufacturing Technology*, 71, 1133–1142, 2014.

[49] M. N. K. Kulkarni, M. S. Patekar, M. T. Bhoskar, M. O. Kulkarni, G. M. Kakandikar, V. M. Nandedkar, Particle swarm optimization applications to mechanical engineering-A review. *Materials Today: Proceedings*, 2(4–5), 2631–2639, 2015.

[50] A. Aggarwal, H. Singh, H., P. Kumar, M. Singh, Multicharacteristic optimization of CNC turned parts using principal component analysis. *International Journal of Machining and Machinability of Materials*, 3(1–2), 208–223. 2008.

[51] G. Kant, K. S. Sangwan, Predictive modeling for power consumption in machining using artificial intelligence techniques. *Procedia CIRP*, 26, 403–407, 2015.

[52] M. H. El-Axir, A method of modeling residual stress distribution in turning for different materials. *International Journal of Machine Tools and Manufacture*, 42(9), 1055–1063, 2002.

[53] X. Wang, C. X. Feng, Development of empirical models for surface roughness prediction in finish turning. *The International Journal of Advanced Manufacturing Technology*, 20, 348–356, 2002.

[54] K. He, M. Gao, Z. Zhao, Soft computing techniques for surface roughness prediction in hard turning: A literature review. *IEEE Access*, 7, 89556–89569, 2019.

[55] P. Muthuswamy, An environment-friendly sustainable machining solution to reduce tool consumption and machining time in face milling using a novel wiper insert. *Materials Today Sustainability*, 100400, 2023.

[56] J. Lu, X. Liao, H. Ou, K. Chen, B. Huang, An effective ABC-SVM approach for surface roughness prediction in manufacturing processes. *Complexity*, (8), 1–13, 2019. https://doi.org/10.1155/2019/3094670

[57] T. T. Nguyen, Q. D. Duong, M. Mia, Multi-response optimization of the actively driven rotary turning for energy efficiency, carbon emissions, and machining quality. *Proceedings of the Institution of Mechanical Engineers, Part B: Journal of Engineering Manufacture*, 235(13), 2155–2173, 2021.

[58] A. Labidi, H. Tebassi, S. Belhadi, R. Khettabi, M. A. Yallese, Cutting conditions modeling and optimization in hard turning using RSM, ANN and desirability function. *Journal of Failure Analysis and Prevention*, 18, 1017–1033, 2018.

[59] K. Safi, M. A. Yallese, S. Belhadi, T. Mabrouki, A. Laouissi, Tool wear, 3D surface topography, and comparative analysis of GRA, MOORA, DEAR, and WASPAS optimization techniques in turning of cold work tool steel. *The International Journal of Advanced Manufacturing Technology*, 121(1–2), 701–721, 2022.

[60] L. G. P. de Souza, J. E. M. Gomes, É. M. Arruda, G. Silva, A. P. de Paiva, J. R. Ferreira, Evaluation of trade-off between cutting time and surface roughness robustness regarding tool wear in hard turning finishing. *The International Journal of Advanced Manufacturing Technology*, 1–32, 3047–3078, 2022.

[61] A. F Torres, F. A. de Almeida, A. P. de Paiva, J. R. Ferreira, P. P. Balestrassi, P. H. da Silva Campos, Impact of stochastic industrial variables on the cost optimization of AISI 52100 hardened-steel turning process. *The International Journal of Advanced Manufacturing Technology*, 104, 4331–4340, 2019.

[62] A. T. Abbas, A. A. Al-Abduljabbar, I. A. Alnaser, M. F. Aly, I. H. Abdelgaliel, A. Elkaseer, A closer look at precision hard turning of AISI4340: Multi-objective optimization for simultaneous low surface roughness and high productivity. *Materials*, 15(6), 2106, 2022.

[63] M. T. Ahmed, H. Juberi, M. M. Bari, M. A. Rahman, A. Rahman, M. A. Arefin, I. Vlachos N. Quader, Investigation of the effect of vibration in the multi-objective optimization of dry turning of hardened steel. *International Journal of International Journal of Industrial Engineering and Operations Management*, 5(1), 26–53, 2023. https://doi.org/10.1108/IJIEOM-11-2022-0059

[64] A. R. Motorcu, The optimization of machining parameters using the Taguchi method for surface roughness of AISI 8660 hardened alloy steel. *Journal of Mechanical Engineering*, 56(6), 391–401, 2010.

[65] L. V. Candioti, M. M. De, M. S. Cámara, H. C. Goicoechea, Experimental design and multiple response optimization. Using the desirability function in analytical methods development. *Talanta*, 124, 123–138, 2014.

[66] S. Raja Bharathi, N. Baskar, Particle swarm optimization technique for determining optimal machining parameters of different work piece materials in turning operation. *The International Journal of Advanced Manufacturing Technology*, 54, 445–463, 2011.

[67] L. Cai, Y. Feng, Y. T. Lu, Y. F. Lin, T. P. Hung, F. C. Hsu, S. Y. Liang, Analytical model for temperature prediction in milling AISI D2 with minimum quantity lubrication. *Metals*, 12(4), 697, 2022.

[68] X. Wang, W. Wang, Y. Huang, N. Nguyen, K. Krishnakumar, Design of neural network-based estimator for tool wear modeling in hard turning. *Journal of Intelligent Manufacturing*, 19, 383–396, 2008.

[69] D. Cica, H. Caliskan, P. Panjan, D. Kramar, Multi-objective optimization of hard milling using Taguchi based grey relational analysis. *Tehnički vjesnik*, 27(2), 513–519, 2020.

[70] A. Iqbal, H. Ning, I. Khan, L. Liang, N. U. Dar, Modeling the effects of cutting parameters in MQL-employed finish hard-milling process using D-optimal method. *Journal of Materials Processing Technology*, 199(1–3), 379–390, 2008.

[71] F. Rabiei, A. R. Rahimi, M. J. Hadad, M. Ashrafijou, Performance improvement of minimum quantity lubrication (MQL) technique in surface grinding by modeling and optimization. *Journal of Cleaner Production*, 86, 447–460, 2015.

[72] H. Aouici, M. A. Yallese, B. Fnides, K. Chaoui, T. Mabrouki, Modeling and optimization of hard turning of X38CrMoV5-1 steel with CBN tool: Machining parameters effects on flank wear and surface roughness. *Journal of Mechanical Science and Technology*, 25, 2843–2851, 2011.

[73] M. K. Dikshit, V. K. Pathak, R. Agrawal, K. K. Saxena, D. Buddhi, V. Malik, Experimental study on the surface roughness and optimization of cutting parameters in the hard turning using biocompatible TiAlN-coated and uncoated carbide inserts. *Surface Review and Letters*, 2022 https://doi.org/10.1142/S0218625X23400024

[74] M. Rafighi, M. Özdemir, A. Şahinoğlu, R. Kumar, S. R. Das, Experimental assessment and TOPSIS optimization of cutting force, surface roughness, and sound intensity in hard turning of AISI 52100 steel. *Surface Review and Letters*, 29(11), 2250150, 2022.

[75] D. Rath, S. Panda, A. Mishra, K. Pal, Particle swarm optimization and machinability aspects during turning of hardened D3 steel. *Journal of Advanced Manufacturing Systems*, 19(04), 641–662, 2020.

[76] J. S. Dureja, V. K. Gupta, V. S. Sharma, M. Dogra, Design optimization of cutting conditions and analysis of their effect on tool wear and surface roughness during hard turning of AISI-H11 steel with a coated-mixed ceramic tool. *Proceedings of the Institution of Mechanical Engineers, Part B: Journal of Engineering Manufacture*, 223(11), 1441–1453, 2009.

[77] H. Aouici, M. A. Yallese, K. Chaoui, T. Mabrouki, J. F. Rigal, Analysis of surface roughness and cutting force components in hard turning with CBN tool: Prediction model and cutting conditions optimization. *Measurement*, 45(3), 344–353, 2012.

[78] M. A. Sofuoğlu, R. A. Arapoğlu, S. Orak Multi-objective optimization of turning operation using hybrid decision making analysis. Anadolu University Journal of Science and Technology. A : Applied Sciences and Engineering, 18(3), 595–610, 2017. https://doi.org/10.18038/aubtda.287801

[79] M. Kaladhar, K. V. Subbaiah, Ch. S. Rao, Simultaneous optimization of multiple responses in turning operations. *Proceedings of the Institution of Mechanical Engineers, Part B: Journal of Engineering Manufacture*, 228(7), 707–714, 2014.

[80] L. R. Bhandarkar, P. P. Mohanty, S. K. Sarangi, Experimental study and multi-objective optimization of process parameters during turning of 100Cr6 using C-type advanced coated tools. *Proceedings of the Institution of Mechanical Engineers, Part C: Journal of Mechanical Engineering Science*, 235(24), 7634–7654, 2021.

[81] S. Padhan, L. Dash, S. K. Behera, S. R. Das, Modeling and optimization of power consumption for economic analysis, energy-saving carbon footprint analysis, and sustainability assessment in finish hard turning under graphene nanoparticle–assisted minimum quantity lubrication. *Process Integration and Optimization for Sustainability*, 4, 445–463, 2020.

[82] X. Cui, J. Guo, Identification of the optimum cutting parameters in intermittent hard turning with specific cutting energy, damage equivalent stress, and surface roughness considered. *The International Journal of Advanced Manufacturing Technology*, 96, 4281–4293, 2018.

[83] B. Fnides, H. Aouici, M. Elbah, S. Boutabba, L. Boulanouar, Comparison between mixed ceramic and reinforced ceramic tools in terms of cutting force components modelling and optimization when machining hardened steel AISI 4140 (60 HRC). *Mechanics & Industry*, 16(6), 609, 2015.

[84] G. Zhou, Q. Lu, Z. Xiao, C. Zhou, C. Tian, Cutting parameter optimization for machining operations considering carbon emissions. *Journal of Cleaner Production*, 208, 937–950, 2019.

[85] S. A. Bagaber, A. R. Yusoff, Energy and cost integration for multi-objective Optimizationn in a sustainable turning process. *Measurement*, 136, 795–810, 2019.

[86] H. Tebassi, M. A. Yallese, S. Belhadi, Optimization and Machinability assessment at the optimal solutions across Taguchi OA, GRA, and BBD: An overall view. *Arabian Journal for Science and Engineering*, 1–29, 2023.

[87] D. Umbrello, G. Ambrogio, L. Filice, R. Shivpuri, A hybrid finite element method–artificial neural network approach for predicting residual stresses and the optimal cutting conditions during hard turning of AISI 52100 bearing steel. *Materials & Design*, 29(4), 873–883. 2008.

[88] A. Hafiz, Prediction of tool life in end milling of hardened steel AISI D2. *European Journal of Scientific Research*, 21(4), 592–602, 2008.

[89] M. Mia, M. H. Razi, I. Ahmad, R. Mostafa, S. M. Rahman, D. H. Ahmed, P. R. Dey, N. R. Dhar, Effect of time-controlled MQL pulsing on surface roughness in hard turning by statistical analysis and artificial neural network. *The International Journal of Advanced Manufacturing Technology*, 91, 3211–3223, 2017.

[90] Z. Jiang, F. Zhou, H. Zhang, Y. Wang, J. W. Sutherland, Optimization of machining parameters considering minimum cutting fluid consumption. *Journal of Cleaner Production*, 108, 183–191, 2015.

[91] S. Bhattacharya, P. P. Das, P. Chatterjee, S. Chakraborty, Prediction of responses in a sustainable dry turning operation: A comparative analysis. *Mathematical Problems in Engineering,* 2021, 1–15, 2021.

[92] R. Chudy, W. Grzesik, K. Zak, Influence of machining conditions on the energy consumption and productivity in finish hard turning. In Hamrol, A., Ciszak, O., Legutko, S., Jurczyk, M. (eds.), *Advances in Manufacturing. Lecture Notes in Mechanical Engineering*, Springer, Cham, 697–705, 2018. https://doi.org/10.1007/978-3-319-68619-6_67

[93] Y. Guo, J. Loenders, J. Duflou, B. Lauwers, Optimization of energy consumption and surface quality in finish turning. *Procedia CIRP*, 1, 512–517, 2012.

[94] S. K. Rajbongshi, D. K. Sarma, A comparative study in prediction of surface roughness and flank wear using artificial neural network and response surface methodology method during hard turning in dry and forced air-cooling condition. *International Journal of Machining and Machinability of Materials*, 21(5–6), 390–436, 2019.

[95] N. Bhone, N. Diwakar, S. S. Chinchanikar, Multi-response optimization for AISI M7 hard turning Using the utility concept. *The Scientific Temper*, 14(1), 142–149, 2023.

[96] G. B. Lee, O. Badrul, Optimization for sustainable manufacturing based on axiomatic design principles: a case study of machining processes. *Advances in Production Engineering & Management*, 9(1), 31, 2014.

Index

A

Adaptive neurofuzzy inference system (ANFIS), 189
Air cooling, 31–33
Analysis of variance (ANOVA), 24, 181, 185–187
Analytic hierarchy process (AHP), 41
Artificial bee colony approach (ABC), 182
Artificial neural network (ANN), 15, 35, 178, 180, 188–189

B

Built-up-edge (BUE), 18, 33, 48, 73, 148

C

Carbide, 3, 5–7, 14–15, 17, 50–51, 56, 58
Carbon emissions (CO_2 emission), 7–10, 20, 26, 29–30, 35, 78, 80, 121
Carbon
  footprints, 82–83, 88
  nanofibre (CNF), 149, 153
  nanotubes (CNTs), 136, 139, 148, 157
Ceramics, 1, 3, 77
Characterization, 137
Chemical vapor deposition (CVD), 136, 183
Compressed air (CA), 20
Computational methods, 180
Conventional cutting fluid (CCF), 120
Cryogenic
  coolant, 99–100, 102
  cooling, 22, 81, 89, 100, 104–107, 113–116, 122–123, 148
  machining 26, 80, 82, 99, 100, 112, 117–118, 123
Cryo
  milling, 106
  minimal quantity lubrication (CMQL), 119
  MQL, 40–41, 83, 120
Cubic boron nitride (CBN), 1–3, 84, 109, 123, 185, 189–190

D

Desirability function (DF), 182, 187
Diamond-like carbon (DLC), 44
Dry
  cutting, 13, 20, 53, 74, 87
  machining, 11–14, 17, 23, 71, 74, 76, 79, 124

E

Economic sustainability, 122
Economical aspects, 9, 23
Electrical discharge machining (EDM), 44, 56
Electrostatic minimum quantity lubrication (EMQL), 29, 30
Energy consumption, 22, 25
Environmental
  aspects, 9, 60
  sustainability, 92, 122
Environmentally conscious machining, 11–12

F

Finite element method (FEM), 75, 188
Flank wear, 17
Focused ion beam (FIB), 44
Fuzzy logic, 180

G

Genetic algorithm (GA), 19–20, 180, 189
Gilbert's approach, 15
Gray relational analysis (GRA), 183–184
Green machining, 11, 60
Grinding, 1, 2, 5

H

Hard
  machining, 1, 2, 4, 5, 14, 19, 40, 58, 60, 75, 99, 131, 133, 162, 174
  turning, 1–6, 16, 34, 75, 77–78, 123, 158, 162
High-speed steel (HSS), 53, 56–58
High-speed surface exposure (HSSE), 161
High strength low alloy (HSLA), 87
Hybrid
  cooling, 40, 60
  nanofluids, 86, 146–147, 153–159
  NFMQL (HNFMQL), 158, 161–162

L

Laser surface texturing (LST), 44
Life cycle assessment (LCA), 60, 83–84, 87–89, 123–124, 163
Liquid
  carbon dioxide (LCO_2), 29, 100–104, 116–118, 123–124
  nitrogen (LN_2), 25, 29, 41, 100–104, 107 108, 111, 114–115, 123–124

M

Machinability, 6, 22–23, 26, 29–30, 37, 39, 46, 48, 56, 115, 118, 163
Machine learning (ML), 177
Manufacturing, 1, 10, 172, 174, 176, 179
Micro-texture, 44–46, 56
Minimum quantity
  cutting fluid (MQCF), 87
  lubrication (MQL), 4, 11, 17–18, 20, 22, 24, 30, 40, 41, 58, 60, 71, 78, 91, 119, 124, 148–149
MOORA, 183, 186
Multi-walled carbon nanotube (MWCNT), 85, 149, 154–155

N

Nanofluid assisted MQL, 12, 24
Nanofluids, 23–25, 39, 60, 131–135, 137, 140–141, 144, 147–149, 159, 162
Nanoparticles, 131–139, 143, 147, 155
Near-dry machining (NDM), 4, 11, 17–18, 78–79, 91
NFMQL, 20, 21, 24–26, 58, 60, 85–86, 148, 153, 158–160, 162

P

Particle swarm optimization (PSO), 20, 180–182, 184, 186
PCBN, 17, 53, 74, 111
Physical vapor deposition (PVD), 106, 136, 188
Principal component analysis (PCA), 181, 184
Pugh matrix, 11, 19–20, 123, 163

R

Regression, 178, 180, 189
Response surface methodology (RSM), 15, 20, 80, 177, 180, 185–189, 191

S

Scanning electron microscopy (SEM), 137–138, 147, 162
Social
  aspects, 9
  sustainability, 122
Solid lubricants, 33–34
Spray impingement cooling, 11, 23, 34, 37, 39–40, 60
Stability, 137, 143–144, 146, 162
Sustainability, 9, 11, 19, 23
  assessment, 7, 19–21, 43, 60, 80, 83, 120, 159–160
  index, 43, 83, 121, 161
Sustainable
  machining, 10, 11, 13, 25, 60, 74, 80–81, 85, 87, 172–173, 176, 187
  manufacturing, 7–10, 13, 40–41, 53, 71–72, 172–173, 190
  techniques, 12, 81
Support vector machine (SVM), 182
Surface roughness, 4, 15, 22, 25
Synthesis, 134

T

Taguchi, 13, 35, 39, 177, 181, 183–186, 191
Textured cutting tools, 41, 44, 48, 50–53, 56
Thermal conductivity, 139–141, 154, 162
TOPSIS, 41, 183, 185–186
Tool
  life, 14, 15, 18, 180
  wear, 5, 22
Total weighted sustainability index (TWSI), 21
Toxic chemical exposure (TCE), 161
Transmission electron microscopy (TEM), 137–138, 147, 162

V

VIKOR, 186
Viscosity, 141, 146–148

W

WASPAS, 183
Weighted sum approach (WSA), 186
WPCA, 13, 39

Z

Zeta potential, 144–145

For Product Safety Concerns and Information please contact our EU representative GPSR@taylorandfrancis.com
Taylor & Francis Verlag GmbH, Kaufingerstraße 24, 80331 München, Germany

www.ingramcontent.com/pod-product-compliance
Lightning Source LLC
LaVergne TN
LVHW010602110826
845149LV00003B/734
* 9 7 8 1 0 3 2 4 0 3 0 0 7 *